ECOLOGICAL CONSEQUENCES OF INCREASING CROP PRODUCTIVITY

Plant Breeding and Biotic Diversity

ECOLOGICAL CONSEQUENCES OF INCREASING CROP PRODUCTIVITY

Plant Breeding and Biotic Diversity

Edited by
**Anatoly Iv. Opalko, PhD, Larissa I. Weisfeld, PhD,
Sarra A. Bekuzarova, DSc, Nina A. Bome, DSc, and
Gennady E. Zaikov, DSc**

Apple Academic Press

TORONTO   NEW JERSEY

Apple Academic Press Inc.	Apple Academic Press Inc.
3333 Mistwell Crescent	9 Spinnaker Way
Oakville, ON L6L 0A2	Waretown, NJ 08758
Canada	USA

First issued in paperback 2021

©2015 by Apple Academic Press, Inc.
Exclusive worldwide distribution by CRC Press, a member of Taylor & Francis Group

No claim to original U.S. Government works

ISBN 13: 978-1-77463-333-5 (pbk)
ISBN 13: 978-1-77188-012-1 (hbk)

Library of Congress Control Number: 2014945460

Library and Archives Canada Cataloguing in Publication

Ecological consequences of increasing crop productivity: plant breeding and biotic diversity/ edited by Anatoly Iv. Opalko, PhD, Larissa I. Weisfeld, PhD, Sarra A. Bekuzarova, DSc, Nina A. Bome, DSc, and Gennady E. Zaikov, DSc.

Includes bibliographical references and index.
ISBN 978-1-77188-012-1 (bound)
1. Agricultural ecology. 2. Agricultural productivity. 3. Crop yields. 4. Plant breeding.
5. Biodiversity. I. Zaikov, G. E. (Gennadiĭ Efremovich), 1935-, editor
II. Opalko, Anatoly Iv, author, editor III. Weisfeld, Larissa I., author, editor
V. Bekuzarova, Sarra A., author, editor V. Bome, Nina A., author, editor

---

S589.7.E36 2014	577.5'5	C2014-904955-2

Apple Academic Press also publishes its books in a variety of electronic formats. Some content that appears in print may not be available in electronic format. For information about Apple Academic Press products, visit our website at **www.appleacademicpress.com** and the CRC Press website at **www.crcpress.com**

ABOUT THE EDITORS

Anatoly Iv. Opalko, PhD

Anatoly Iv. Opalko, PhD, is Head of the Physiology, Genetics, Plant Breeding and Biotechnology Division at the National Dendrological Park "Sofiyivka" of the National Academy of Sciences of Ukraine, Uman, Cherkassy District, Ukraine, and Professor at Uman National University of Horticulture, Uman, Ukraine, as well as Head of the Cherkassy Regional Branch of Vavilov Society of Geneticists and Breeders of Ukraine. He is also a prolific author, researcher, and lecturer. He has received several awards for his work, including the badge of honor "Excellence in Agricultural Education." He is member of many professional organizations and on the editorial boards of the Ukrainian biological and agricultural science journals.

Larissa I. Weisfeld, PhD

Larissa I. Weisfeld is a senior researcher at the Emanuel Institute of Biochemical Physics, Russian Academy of Sciences, Moscow, Russia; a member of the All-Russia Vavilov Society of Geneticists and Breeders; the author of about 300 publications in scientific journals and several patents and conference proceedings; and the co-author of three publications on new cultivars of winter wheat. Her main field of interest concerns basic problems of chemical mutagenesis, mutational selection, and the mechanism of action of para-amonobensoic acid. She has worked as a scientific editor in the publishing house Nauka (Moscow) and in the journals *Genetics* and *Ontogenesis*.

Sarra A. Bekuzarova, DSc

Sarra A. Bekuzarova, DSc in Agriculture, is Head of the Laboratory of Plant Breeding and Feed Crops of the North Caucasus Institute of mountain and submontane agriculture of the Republic of North Ossetia-Alania, Russia; Professor at the Gorsky State University of Agriculture, Vladikavkaz, Republic of North Ossetia-Alania, Russia, as well as Professor at Kosta Khetagurov North-Ossetia State University, Vladikavkaz, Republic of North Ossetia-Alania, Russia. She is also a prolific author, researcher, and lecturer. She is a corresponding member of the Russian Academy of Natural Sciences. She is a member of the International Academy of Authors of the Scientific Discoveries and Inventions, the International Academy of Sciences of Ecology, Safety of Man and Nature, the All-Russian Academy of Non-traditional and Rare Plants, and the International Academy of Agrarian Education. She is a member of the editorial boards of several scientific journals.

Nina A. Bome, DSc

Nina A. Bome, DSc in Agriculture, is Professor and Head of the Department of Botany, Biotechnology and Landscape Architecture at the Institute of Biology at the Tyumen State University, Tyumen, Russia. She is the author of the monographs, articles, school-books, and patents, and she is a lecturer. She is the Director and Founder of the Scientific School for Young Specialists. She is the author of about 300 publications. She participates in long-term Russian and international programs. Her main field of interest concerns basic problems of adaptive potential of cultivated crops, mutagenesis, possibility of conservation, enhancing biodiversity of plants, methods of evaluation of plants' resistance to the phytopatogens and other unfavorable environmental factors, and genetic resources of cultivated plants in the extreme conditions of the Western Siberia.

Gennady E. Zaikov, DSc

Gennady E. Zaikov, DSc, is Head of the Polymer Division at the N. M. Emanuel Institute of Biochemical Physics, Russian Academy of Sciences, Moscow, Russia, and professor at Moscow State Academy of Fine Chemical Technology, Russia, as well as professor at Kazan National Research Technological University, Kazan, Russia. He is also a prolific author, researcher, and lecturer. He has received several awards for his work, including the the Russian Federation Scholarship for Outstanding Scientists. He has been a member of many professional organizations and on the editorial boards of many international science journals.

CONTENTS

PART I: MOUNTAIN BIOTA OF CULTIVATED MEADOW PASTURE IN NORTH CAUCASUS

PART II: PLANT BREEDING IN THE WESTERN SIBERIA: AGROECOLOGICAL APPROACHES

LIST OF CONTRIBUTORS

S. V. Arsentyev
Post-graduate student at Department of Botany, Biotechnology and Landscape Architecture of Institute of Biology of Tyumen State University, st. Semakova, d. 1, Tyumen, 625003,Russia, tel. +7(3452) 64-07-24, e-mail: arsentev_serge@mail.ru

S. S. Basiev
DSc of the Agriculture, Professor, Head of the Department of Plant Growing, Gorsky State Agrarian University, st. Kirov, d. 37, Vladikavkaz, Republic of North Ossetia Alania, 362040, Russia, tel. (8672) 54-87-17, +7(919) 428-65-26, e-mail: basiev_s@mail.ru

S. A. Bekusarova
DSc of the Agriculture, Honored the Inventor of Russian Federation, Professor, Gorsky State Agrarian University, Republic of North Ossetia Alania, st. Kirov, d. 37, Vladikavkaz, 362040, Russia, tel. (8672) 36-20-40 North-Caucasian Research Institute of Mountain and Foothill Agriculture, Republic of North Ossetia Alania, Suburban region, Mikhailovskoye village, st. Williams, d. 1, 363110, Russia, e-mail: bekos37@mail.ru

V. A. Belyayeva
PhD of the Agriculture, Senior Scientist, Institute for Biomedical Research of Vladikavkaz Scientific Centre of RAS and RNO-Alania Government, st. Pushkinskaya, d. 40, Vladikavkaz, Republic North Ossetia Alania, 362019, Russia, tel. +7(906)-494-44-93, e-mail: pursh@inbox.ru

A. Ja. Bome
PhD of the Agriculture, Senior Research Associate at the Tyumen Basing Point of the All-Russia Research Institute of Plant Growing of N.I. Vavilov, st. Bol'shaya Morskaya, d. 1, Saint Petersburg, 190000, Russia, tel. +7(812) 314-22-34, e-mail: office@vir.nv.ru

N. A. Bome
DSc of the Agriculture, Professor, Head of the Department of Botany and Plant Growing, Professor in Tyumen State University, st. Semakova, d. 1, Tyumen, 625003, Russia, tel. +7(3452) 46-40-61, +7(912) 923-61-77, e-mail: bomena@mail.ru

V. I. Bushuyeva
DSc of the Agriculture, Professor, Associate Professor of the Department of Selection and Genetics at Belorussian State Agricultural Academy, st. Michurin, d. 5, Gorki, 213407, Belarus, tel. +3750223379674, +375296910383, e-mail: vibush@mail.ru

Ya. V. Byelyk
Post-graduate student in Institute of Bioenergy Crops and Sugar Beet NAAS, st. Clinichna, d. 25, Kyiv, 03141,Ukraine, tel. +3(8097) 196-04-97, e-mail: belyar87v@ukr.net

A. V. Doronin
PhD of the Economics, Senior research scientist at National Scientific Centre Institute of the Agrarian Economics of Ukraine, st. Heroiv Oborony, d. 10, Kyiv, 03680, Ukraine, +38 095 805-63-34, e-mail: andredor@meta.ua

V. A. Doronin
DSc of the Agriculture, Professor, Head of Laboratory of the Institute of Bioenergy Crops and Sugar Beet of NAAS, st. Clinichna, d. 25, Kyiv, 03141, Ukraine, tel. +38 067-466-47-15, e-mail: doronin@tdn.kiev.ua, vladimir.doronin@tdn.org.ua

I. A. Dudareva
PhD of the Agriculture, Junior Researcher, Tobolsk Complex, Scientific Station of the Ural Branch of the Russian Academy of Sciences, st. Academician Yuriy Osipov, d. 15b, Tobolsk, 626150, Russia, e-mail: tbsras@rambler.ru

V. I. Gasiv
PhD of the Agriculture, Researcher, North Caucasian Research Institute of Mountain and Foothill Agriculture, Republic of North Ossetia Alania, Suburban region, Mikhailovskoye village, st. Williams, d. 1, 363110, Russia

M. D. Gazdarov
Director of the Republican Children's Ecological and Biological Center, Kosta Avenue, d. 28, Republic of North Ossetia Alania, Vladikavkaz, 362040, Russia, tel.8 (8672) 75-55-68

L. S. Gishkayeva
PhD of the Agriculture, Head of Department of Agronomics, Chechen State University, st. Sheripov, d. 32, Grozny, Chechen Republic, 364907, Russia, +7(8712) 21-20-04

I. M Haniyeva
DSc of the Agriculture, Professor, Head of Department of Plant-grower and Plant-breading, V.M. Kokov's Kabardino-Balkarsky State Agrarian University, Lenin Avenue, d. 1v, Nalchik, Kabardino-Balkaria Republic, 360000, Russia, tel. +7(8662) 47-41-77.

L. M. Karpuk
PhD of the Agriculture, Associated Professor of Bila Tserkva National Agrarian University, sq. Soborna, d. 8/10, Kyiv region, Bila Tserkva, 9100, Ukraine, tel. +38 067 991 91 17, e-mail: zuikes@ukr.net

M. V. Kataeva
Post-graduate student of North Caucasian Mountain Scientific Research Institute of Mountain and Foothill Agriculture, Suburban district, st. Williams, d. 1, village Michajlovskoye, North Ossetia Alania, Vladikavkaz, 363110, Russia, tel. +7(8672) 530708.

N. N. Kolokolova
PhD of the Biology, Associate Professor of the of the Agriculture for Botany, Biotechnology and Landscape Architecture of Tyumen State University, st. Semakova, d. 1, Tyumen, 625003, Russia, tel. +7(3452) 64-07-24

M. O. Korneeva
PhD of the Biology, Leading Researcher, Institute of Bioenergy Crops and Sugar Beet of National Academy of Agrarian Science of Ukraine, st. Klinichna, d. 25, Kyiv, 03141, Ukraine, +380 675 96 08 72, e-mail: mira31@ukr.net

I. S. Kosenko
DSc of the Biology, Professor, Director of the National Dendrological Park "Sofiyivka" of NAS of Ukraine, st. Kyivska, d. 12/a, Uman, Cherkassy region, 20300, Ukraine, e-mail:sofievka@ck.ukrtel.net

D. P. Kozaeva
Post-graduate student of the Mountain and Foothill Agriculture, Republic of North Ossetia Alania, Suburban region, Mikhailovskoye village, st. Williams, d. 1, 363110, Russia, tel. (8672) 53-07-08.

N. M. Kucher
Master of Science in Agriculture, Junior Researcher of the Physiology, Genetics, Plant Breeding and Biotechnology Division at National Dendrological Park "**Sofiyivka**" of NAS of Ukraine, st. Kyivska, d. 12-a, Uman, Cherkassy region, 20300, Ukraine, tel. +380974894321, e-mail: natalochka_sof@ukr.net

Insa Kuehling
Post-graduate student at University of Applied Sciences Osnabrueck, P.O. Box 1940, Osnabrueck, 49009, Germany +49 170 2046755, e-mail: i.kuehling@hs-osnabrueck.de

A. N. Michailova
Assistant at Department of Botany, Biotechnology and Landscape Architecture of Institute of Biology of Tyumen State University, st. Pyrogov, d. 3, Tyumen, 625043, Russia, tel. +7(3452) 64-07-24, e-mail: ma_noy@mail.ru

M. M. Nenka
Post-graduate student, Junior Researcher of Institute of Bioenergy Crops and Sugar Beet of National Academy of Agrarian Science of Ukraine, st. Klinichna, d. 25, 03141, Ukraine, tel. +380501024848, e-mail: nenka88@i.ua

A. I. Opalko
PhD of the Agriculture, Associate Professor, Head of the Physiology, Genetics, plant Breeding and Biotechnology Division in National Dendrological Park "Sofiyivka" of NAS of Ukraine, st. Kyivska, d. 12-a, 20300, Uman, Cherkassy region, 20300, Ukraine; and Professor of the Genetics, Plant Breeding and Biotechnology Chair in Uman National University of Horticulture, st. Instytutska, 1, 20305, Uman, Cherkassy region, Ukraine, tel. +380506116881, e-mail: opalko_a@ukr.net

O. A. Opalko
PhD of the Agriculture, Associate Professor, Senior Researcher of the Physiology, Genetics, Plant Breeding and Biotechnology Division in National Dendrological Park "Sofiyivka" of NAS of Ukraine, st. Kyivska, d. 12-a, Uman, Cherkassy region, 20300, Ukraine, +380506116881, e-mail: opalko_o@ukr.net

V. V. Polishchuk
PhD of the Agriculture, Docent of Uman National University of Horticulture, Head of Department, st. Institutska, 1, Cherkassy region, Uman, 20305, Ukraine, tel. +38 098 502 03 70, e-mail: pol.val@i.ua

E. I. Ripberger
Graduate student at Institute of Biology of the Tyumen State University, st. Semakova, d. 1, Tyumen, 625003, Russia. +7(3452) 46-40-61, +491763-544-58-14, e-mail: lena-umka@yandex.ru

S. G. Shichova
Student at Department of Botany, Biotechnology and Landscape Architecture of Institute of Biology of Tyumen State University, st. Pirogov, d. 3, Tyumen, 625043, Russia, tel. +7(3452) 64-07-24.

Dieter Trautz
Dr. sc. agr., Professor of Ecology at University of Applied Sciences of Osnabrueck, P.O. Box 1940, 49009 Osnabrueck, Germany, tel. +49(0)541 969-5058, fax +49(0)541 969-5201, e-mail: d.trautz@hs-osnabrueck.de

Ju. B. Trofimova
PhD of the Agriculture, Leading researcher of Tyumen Research Institute Cryology Resources Tyumen State Oil and Gas University, st. Volodarskii, d. 56, Tyumen, 625000, Russia, tel. 8(3452) 46 39 78, E-mail: trof_jb@mail.ru

E. A. Tyumentseva
Assistant at Tyumen State University, st. Pyrogov, d. 3, Tyumen, 625043, Russia, tel. +7(3452) 64-07-24, e-mail: tumenea@mail.ru

S. P. Vasilkivskiy
DSc of the Agriculture, Professor, Head of the Genetic, Breeding and Seed Production Chair at the Bila Tserkva National Agrarian University, sq. Soborna, d. 8/1, Kyiv region, Bila Tserkva, 09117, Ukraina, +380988951502, e-mail: vasilsp@gmail.com,

L. I. Weisfeld
Senior Research at N. M. Emanuel Institute of Biochemical Physics RAS, st. Kosygin, d. 4, Moscow, 119334, Russia, tel. +7 (916) 227-86-85, e-mail: liv11@yandex.ru

R. A. Yakymchuk

PhD of the Agriculture, Associate professor, Head of the Natural Sciences Faculty at Pavlo Tychyna Uman State Pedagogical University, st. Sadova, d. 2, Uman, Cherkassy region, 20300, Ukraine, e-mail: peoplenature@rambler.ru

A. I. Yurchenko

Ph.D. of the Agriculture, Assistant Professor of the Genetics, Breeding and Seed Production Chair at the Bila Tserkva National Agrarian University, Sq. Soborna, d. 8/1, Bila Tserkva, Kyiv region, 09117, Ukraine, tel. +380988951502, e-mail: anatoliy.y@gmail.com

I. L. Zamorska

Ph.D of the Agriculture, Associate professor of the Technology of Storage, Preservation and Processing of Fruits and Vegetables Chair at Uman National University of Horticulture, st. Instytutska, d. 1., Uman, Cherkassy region, 20305, Ukraine, e-mail:zil1976@mail.ru

V. V. Zamorskyi

DSc of the Agriculture, Professor, Deputy Rector Uman of the National University of Horticulture, st. Instytutska d. 1, Uman, Cherkassy region, 20305, Ukraine, e-mail: zvv55@mail.ru

N. M. Zapolska

Ph.D of the Agriculture, Assistant Prifessor of the Genetic, Breeding and Seed Prodection at Breeding Institute of the Bioenergy Crops and Sugar Beet of NAAS, st. Clinichna, d. 25, Kyiv, 03141, Ukraine, tel. +38 066 733-35-75, e-mail: zapolska_katerina@i.ua

LIST OF ABBREVIATIONS

CHEMICALS

DAB	1,4-bisdiazoacetylbutan
DES	diethyl sulfat
DNS	dimethyl sulfat
EI	ethylene imine
NDMU	N-nitrozodymethyl urea
NEU	N-nitrozo-N-ethyl urea
NMB	nitrozomethylbiuret
NMU	N-nitozo-N-methyl urea

OTHERS

ANOVA	ANalysis Of VAriance
Al^{+++}	ion of aluminum
BS	binder of sterility
°C	°C - degree Celsius
C_c	chemical substance concentrations
Ca^{++}	ion of calcium
CIS	Commonwealth of Independent States
CMS	cytoplasmic male sterility
CV	coefficient of variation.
cwt	centner enxybnm
GVTS	Gorki variety testing station
DB	digestible protein
DSTU	state standards of Ukraine
U	the European Union
F_1, F_2, F_3, F_4 ...	generations of organisms from first to fourth et ctr
g	gram
H^+	ion of hydrogen
ha	hectare – is area unit that equal to 10,000 square meters in the measure metric system. In Ukraine hectare is the main unit of land area
HTC	hydrothermal potential

IBCA	increased background of fertilizing — the common area;
IBEA	increased background — the extended area of supply
kg	kilogram - is mass unit in the International System of Units (CI) and some other metric systems
km	kilometer, kilometre
l	liter - volume metric unit that equal to 1 cubic decimeter
LSD	small statistical distinction
M	gram-molecule
$M_1 - M_7$	generations of mutants from first to seventh
mg	milligram
Mg^{++}	ion magnesium
ml	milliliter
mm	millimeter, millimeter
mm per year	millimeters per year - atmospheric fallouts amount
MS	male sterility
n	haploid chromosomal set, chromosome complement
2n	diploid
NAAS	National Academy of Agrarian Sciences of Ukraine
NAS	National Academy of Science
NBCA	the common area of supply
NBCA	normal background of fertilizing — the common area of supply
NBEA	normal background — the extended area of supply
NDP	National dendrological park "Sofiyivka" of NAS of Ukraine
NFE	nitrogen free extractives
NSC	National Science Centre
pH	the hydrogen ion concentration
P, mm	precipitation amount
R	Roentgen
RED	Renewable Energy Directive
SEG	Society of Exploration Geophysicists
sm	centimeter
sobole	a shoot, stolon or sucker, a creeping underground stem that produces roots and buds.
SSH	simple sterile hybrids

t	ton - is mass unit, that equal to 1000 pounds in the units technical system
(T, °C)	air temperature
th. t	thousand tones
TPS	Thermal power station
UAH	hryvnia - the national currency of Ukraine
UPOV	The International Union for the Protection of New Varieties of Plants in Russian - variant distinctiveness, homogeneity, stability = in English DUS - distinctness, uniformity, stability
urc	units of regeneration coefficient
USA	the United States of America
US $, $ USD	the United States dollar
USD/t $/t	the United States dollar per ton
$/l	The United States dollar per liter
SSH	simple sterile hybrids
VAT	Value Added Tax - indirect tax that is accrued and paid according to the Title V of the Tax Code of Ukraine norms
VIR	All-Russian Institute of Plant Growing
X-rays	X-ray exposure, X-ray irradiation, roentgenization, X-raying
y-axis	acsis of ordinates

LIST OF SYMBOLS

$\sum P$	precipitation amount
$\sum T$	sum of temperatures higher than +10°C for some period of time
a_i	corresponding number of ill seedlings
HTC	hydrothermal coefficient
I	degree of damage
k	maximum degree of damage
n_1	number of days after notching was done to the appearance of the first signs of callus
n_2	number of days after notching was done to the completion or termination of callus development
N	amount of plants in a test
R	regeneration coefficient, urc
S	intensity of callus genesis, points

PREFACE

As the history of our planet shows, living organisms, including humans—and now foremost humans, affect the environment, which is undergoing more and more changes due to increasing anthropogenic pressure. Other plants and organisms are also affected indirectly because of such changes.

The excessive influence of *Homo sapiens* and his "flatmates" from the same environment have had threatening consequences to the survival of humankind itself since the beginning of the third epoch of globalization, when for every 20 minutes one species disappeared, which was 1,000 times faster than for most of the history of our planet. Similar processes also concern atmospheric, water, soil mineral, and other natural resources. Consequently, an extremely complicated and vulnerable ecosystem, which formed and gradually evolved during thousands of centuries, is now being destroyed by the irrational human activities during the past few years, months, or even days. Taking into account the inherent egoism of each human individually, as well as the group egoism of each state or almost any ethnic, political, clerical, or other entity, it is hard to expect any voluntary self-restriction, however considerable, even conditioned by real anxiety of exhaustion, pollution, or destruction of soil or any other natural resources.

Real instruments of stimulation of environmental protection measures are still not available in most of the countries of our planet.

Reasonable use of the progress in selection can contribute to the slowing and in prospects even reversion of processes of destruction of soil resources.

The growth of food production in the world during the past 50 years was allowed mainly by the use of high-yielding varieties of agricultural crops, application of fertilizers and pesticides; irrigation; and other components of the intensification of plant cultivation, which have absolutely become necessary now. It is necessary to admit that the progress in selection is the key component of efficient land use among all mentioned and many other components of innovative plant cultivation.

The increase of productivity of the main food cultures up to 1.5 times by 2050 due to the selection is foretold. This will allow the livelihood of further nine billion population of earth without considerable increase of plough land, which is extremely important. Creation of new varieties not only adapted to the standard natural changes of growth conditions, but it is also capable of providing stably expected productivity and yielding quality in the environment modified by human activity is the aim of new selection programs. It is already assumed that the universal application of such varieties can improve the ecological situation without infringing upon the interests of both producers and consumers of agricultural production across the world, as well as in each particular country.

Namely, these reasons induced the compilers of this book to contribute reports of different leading scientists in agricultural biology and young researchers from different countries, discussing the ways of improvement of cultivated plants, methodical questions of different methods of inducing hereditary variability, and means of evaluation of material under selection. A special place is attributed to the studies, concerning the conservation and restoration of biotic diversity, the search of alternative energy sources, and other issues of similar importance related with the theoretical preconditions of sustainable agricultural production of the third millennium.

INTRODUCTION

All kinds of modern anthropogenic activities, including agricultural production, cause a rapid increase in pollution. In this regard, permanent reduction in the quality of food in some cases can be dangerous. Agricultural science is seeking safe methods for obtaining of the environmentally friendly products, offering different ways of improving the biotic diversity of cultivated plants and diversification of their cultivation. One of the major components of biodiversity is the genetic resource of cultivated plants and their wild relatives.

This book offers various methods of agriculture management in ecologically and economically different regions. The book proposes different methods of solving agricultural problems in ecologically and geographically diverse regions, such as Western Siberia, Ukraine, Caucasus, and Belarus. All variants are bound by one idea consisting in the study of ecological danger of agricultural production, seeking in different methods of eradicating this danger. The chapters have been contributed by prominent scientists—experts in the fields of genetics, plant breeding, biology, and agronomics.

PART I: MOUNTAIN BIOTA OF CULTIVATED MEADOW PASTURE IN NORTH CAUCASUS

S.A. Bekusarova studies the adaptability of widely used agricultural crops in the conditions of vertical zonality of North Ossetia, Kabardino-Balkaria, and Chechen Republic. Bekusarova's team have proposed various methods of creation of economically valuable forms of red clover and other forage herbs for cultivation in mountain and foothill conditions, as well as selection of native species of clover and Timothy grass, collection of herbaria of valuable samples, the conservation of existing natural and cultural agrocenoses, and restoration of phytocenoses of red clover. Research of the Ukrainian breeders is devoted to traditional directions for the Southwest region of the former USSR, that is, improvement of production of sugar beet and fruit plants, in particular apple, pear, filbert, and strawberry as well as Timothy grass and amaranth.

S.S. Bassiev et al. have studied the features of varietal potatoesunder different conditions of planting in foothill zone and efficiency of mineral nutrition of potatoes.

PART II: PLANT BREEDING IN THE WESTERN SIBERIA: AGROECOLOGICAL APPROACHES

The valuable forms of medical, food, forage, and technical plants are concentrated in Western Siberia. However, subzero low bioclimatic potential of the region and development of oil-production complex industry and urbanization really diminish the abilities of different kinds of species, and sometimes conduce leading to their complete disappearance.

Bome et al. have studied the ecological and biological potential of cultural plants in the Institute of Biology of Tyumen State University and in his the basing point of the All-Russia Research Institute of Plant Breeding, which solvedthe following tasks: creation and addition to collections of plants and microorganisms from natural populations and their introduction from scientific institutions; formation of bank of plant seeds and clean cultures of phytopathogenic fungi in order to preserve the gene pool; and research of basic properties of sod-podzolic soil in the north of Tyumen region.

It is shown that environmental factors can be used as a proxy for field germination of seeds and of the biological stability of the plants during the growing season. It is found that the adaptive properties of seeds during germination and plant development in ontogenesis depend on the genotype and meteorological factors. It is also found that the varieties of spring wheat are more sensitive to the limiting factors of environment than hybrids, derived by crossing these varieties. The relationship between soybean cultivars, chlorophyll meter readings, and Rhizobia inoculation in Western Siberia was derived.

PART III: PLANT BREEDING IN THE UKRAINE: AGROECOLOGICAL APPROACHES

Research of the Ukrainian breeders is devoted to traditional directions for the South-west region of the former USSR, namely improvement of production of sugar beet and fruit plants, in particular apple, pear, filbert, and strawberry. Thus, the main attention is on increasing the adaptive potential of the cultivated plants as a leading factor of modern agrarian production. Post-traumatic regeneration potential of pear cultivars and species of the genus *Pyrus* L. was studied. The suggestion is made that a regeneration ability indicator confirms indirectly the level of ecological adaptation of the genotypes under study, and the periods of the highest regeneration activity can be favorable for vegetative propagation, including propagation by cutting and grafting *in vitro* and other technological processes causing plant damage.

Breeders of sugar beet seek different ways of improving the efficiency of sugar production, including creation of new genotypes, improvement of technologies of their seed material processing, and assessment of resistance to major pests of beet.

Taking into consideration the necessity of reduction of energetic dependence, scientists studied the possibilities of bioethanol production from plant-breeding products, in particular from a sugar beet, which is important not only to Ukraine but also to many other states—importers of oil. The study of cytogenetic aftermath of the soil contamination by mutagens evoked by TPS emission in winter wheat plants evaluates the risks to other higher organisms including humans.

PART IV: ROLE OF CHEMICAL MUTAGENESIS IN INDUCTION OF BIODIVERSITY OF AGRICULTURAL PLANTS

Rapoport, a well-known scientist in the scientific community, discovered the theoretical and practical bases of methods of chemical mutagenesis. A method has been applied until now for the creation of new varieties of different cultures, for example, wheat, barley, and vegetables. In Ukraine, under S.P. Vasilkovsky's leadership, varieties of mutant lines of winter wheat were created; the chromosome aberrations of induced mutants and the electrophoretic spectra of gliadin in polyacrylamide gel were studied. Spectra of gliadin give the opportunity to identify mutant lines.

At the end of the division, the cytogenetic analysis of the effect of chemical mutagen in the cells of meristem of *Crepis capillaris* plants is reported. The cytogenetic mechanism of induction of rearrangements is researched under the effect of mutagens.

PART V: SELECTION OF LEGUMINOUS HERBAGES IN BELARUS

This division describes the achievements in the breeding of legumes in Belarus (V.I. Bushuyeva, Belorussian State Agricultural Academy). The division studies the breeding of *Galega orientalis* and red clover, the most significant forage legumes for agricultural production, which have a significant impact on the preservation and restoration of soil fertility, saving of energy and labor resources, and qualitative improvement of the environment. At the same time, these are the most effective source of cheap highly nutritious livestock feed. One hectare without the application of mineral nitrogen can produce at least 100 metric canters of feed units and 15–17 metric canters of digestible protein. Moreover, each hectare of legumes accumulates in the soil 180–220 kg of biological nitrogen on average, which does not pollute the environment unlike the mineral nitrogen. When legumes are sown as a predecessor for barley, wheat, and triticale, the grain yield further increases by 8–10 t/ha, and its protein content increases by 1.5–2.0 percent.

— **Larissa I. Weisfeld, Anatoly Iv. Opalko, and Gennady E. Zaikov**

PART I

MOUNTAIN BIOTA OF CULTIVATED MEADOW PASTURE IN NORTH CAUCASUS

CHAPTER 1

NATIVE SPECIES OF CLOVER AND TIMOTHY-GRASS IN MOUNTAIN PHYTOCENOSIS: APPLICATION IN SELECTION FOR THE CREATION OF ECONOMIC VALUABLE FORMS

SARRA AB. BEKUZAROVA

CONTENTS

1.1 INTRODUCTION

The most important chain of the adaptive approach in selection is the elaboration of principles and methods of phytocenotic selection, that is, the creation of competitive cultivars capable of adapting in mixed crops in the meadow diversity of mountain hayfields and pastures.

For the sustainable development of mountain phytocenosis biodiversity, it is necessary to set up systems of environmentally differentiated cultivars of fodder crops with tolerance to extreme and destabilizing environmental conditions. Environment-evolutionary principles are becoming dominant in the selection strategy of fodder crops in recent years. They are based on the theory of adaptive system of growing plants and provide for the creation of geographically and environmentally differentiated cultivars [1].

As it is known [2, 3], wild plants from natural ecosystems have such valuable traits as longevity, frost hardiness, drought resistance, and high concentration of nutrient elements.

The existing cultivars of meadow-pasture grasses, as practice shows, are unable to form agrocenosis in specific mountain conditions, because they do not have enough productive longevity. Recommended cultivars of red clover (*Trifolium pratense*) are not efficient in undersow due to low competitive capacity with native species of legumes. They have low survival rate of shoots, and even those plants that have survived do not live for long and soon drop out of the grass stand, which leads to the excessive labor and capital costs [4, 5].

For mountain regions with complex environmental conditions, cultivars with resistance to stress are needed. Such cultivars are lacking nowadays, because in Russia the selection for potential productivity is traditionally preferred. A combination of high productivity and environmental resistance is a hard task. Decrease of adaptation level of modern cultivars is the result of limitation of their genetic base due to the use of a small number of genotypes as well as long and intensive selection in the constantly repeating environmental conditions.

For the long period of selection studies, we have defined that creation of most productive cultivars in mountain environmental zone by well-known effective methods which is unavailable for the cultivars of hayfield-pasture type, because created populations have a major drawback—the low adaptive capacity in conditions of vertical zonal differentiation of mountain slopes. Besides, studied selection samples in one-species crops had a minimal competitive capacity. All cultivars in selection are tested in one-species crops. We suggest using mixed crops in the evaluation of competitive capacity.

For the creation of competitive cultivars for mountains, phytocenosis selection samples were evaluated in the mountains (600; 900; 1,200; and 2,000 m above sea level), sowing selected plants in the mixture with grasses and motley grasses of the wild flora.

Among gramineous plants, we have chosen timothy-grass (*Phleum pratense*), because it belongs to loose-bush plants, which have the node of bushing out on the small depth (1–5 cm). The loose-bush gramineous plants (*P. pratense, Festuca pratensis, Dactylis glomerata, Arrhenatherum elatius*) have overground shoots coming out from one node of bushing under the acute angle to the main shoot, forming as a result of a loose bush. Every year new shoots grow in this bush, and each has its own node of bushing out. In their turn, new shoots grow from those nodes, and the bush increases in its volume, but it remains loose, because new shoots after coming out of the ground stand not far from each other. Loose-bush gramineous plants have denser root rosette than rhizome species.

The choice of gramineous loose-bush component is based on its capacity to form the dense sod and to replace legumes. This biological peculiarity of the loose-bush gramineous plant provides for a possibility to evaluate the selection sample of clover in the tough conditions of a phytocenosis.

Gramineous and legume grasses have different demand for warmth, light, and nutrients. Legumes absorb larger amount of calcium, manganese, and chlorine from soil, while the gramineous plants exceed legumes in the uptake of phosphorus and silicon. Calcium and chlorine are present mainly in the lower layers of soil with limestone streaks coming from the mother rock, and legume grasses with long roots can better absorb them from these layers. Due to nodule bacteria living on legume roots, the gramineous plants are better supplied with nitrogen. At the expense of their dying out roots, they supply the components of gramineous grass stands with nitrogen nutrition.

The process of bushing out in gramineous plants usually begins 1–1.5 months after the emergence of shoots over ground. The formation of shoots happens at the expense of photosynthesis in the green parts of the plants and not at the expense of reserve substances. In natural pasture plant communities, the loose-bush gramineous plants using their sod-forming capacity can replace the legume component (especially red clover, bird's foot trefoil, sainfoin). Thus, for the evaluation of their competitive ability, legumes should be sown with such gramineous plants. This biological trait of gramineous plants provides a possibility to evaluate the legume component in tough conditions.

Evaluation of legumes in grass mixtures is fulfilled on the quantity of both shoots per plant in the first year of life and survived plants per square measure throughout the trial. If in the pure crop the number of shoots reaches 12–15 dependent on the sample, in the mixture with gramineous plants the number of shoots per plant does not exceed 3–5. The number of flowering shoots per stem reduces from 7–10 in the pure crop to 2–3 in the mixture.

The optimal proportion of legume component in natural conditions of an ecosystem (i.e. under ideal ratio of grasses on pastures) should be 40–50 percent [2].

If the sample has withstood the test in comparison with the standard as a pure crop and as an individual plant on a complex set of traits, but received low marks for competitiveness (9-grade international system), it will be classified unfit for the formation of a meadow-pasture cultivar. However, it can be used as a basic material for cultivars of field fodder production.

Evaluation of samples in phytocenotic selection on competitiveness includes registration of number of legume shoots, height of plants, number of generative organs, and number of seeds in inflorescence. According to the procedure adopted by CMEA countries [6–8], the competition capacity in grass stands is determined by the following 9-point system:

1—very bad, when clover plants are depressed, develop poorly and fall out from grass stand the next year;

3—bad, when after weak growth plants have few shoots on the second and third years;

5—medium, when more than half of the plants develop well and flower in the year of sowing;

7—good, when 70 percent of legume plants survive on second and third years with good bushiness (4–5 shoots) and flowering inflorescences;

9—very good with survival of all sown legume plants.

But legumes, including clover, in natural ecosystems usually compete with motley grasses, which are dominant in the grass stand (above 50%). Due to this competition many legume species drop out of the grass stand.

Newly created cultivars of meadow-pasture legume grasses undersown in mountain phytocenosis have low adaptive capacity. Thus, the binary mixture with the gramineous component is not efficient enough for the selection evaluation of samples.

The aim of this study is the selection of the most adaptive plants for the complex traits, among which the main aim is the competitiveness in the grass mixture including gramineous component (timothy-grass) and motley grass representative (*Poterium polygamum*).

1.2   MATERIALS AND METHODOLOGY

To achieve the mentioned aim clover and timothy-grass were sown in ratio 1:2 (one part of clover and two parts of timothy). *P. polygamum* was added in the amount of 15–20 percent of clover–timothy mixture. Competitive samples of legumes were selected in the second year. Those samples were considered to be competitive, which survived. On their base a new meadow-pasture cultivar was formed.

P. polygamum belongs to family *Rosaceae*. This plant has high fodder qualities, a taproot, high winter hardiness, longevity, and resistance to cold and drought. For such qualities, it is included into the mixture of motley grasses. In the year of sowing, *P. polygamum* develops a vigorous root system and a rosette of leaves. This helps to determine the most productive and competitive plants in the early period of development of selection samples in collection nurseries. In the mixture with the legume-gramineous component *P. polygamum* secures the high yield of forage mass.

We have included in the experiment evaluation of 18 samples of meadow clover in the collection nursery. Zoned cultivar of red clover from North Caucasian Region—Daryal—was used as standard. The square of each plot is 5 m^2. The studied samples were compared with the zoned cultivar known for its high longevity and quite stable productivity from year to year. On each plot, we have sown the seed mixture of clover (7.5 g), timothy-grass (4 g), and *P. polygamum* (2.3 g, which consists 20% of the mixture of clover and timothy).

1.3   RESULTS AND DISCUSSION

The results of the experiments are summarized in Table 1.1.

TABLE 1.1 Competitiveness of red clover samples in mixture with timothy-grass and *Poterium polygamum* depending on the duration of growing from cultivar or sample

Name of sample	Number of stems per 1 m² dependent on years of life		Survival of plants of the second year of life in comparison with survival on the first year of life (%)
	First year	Second year	
Daryal (standard)	12.5	4.2	33.6
Alan	15.2	4.8	31.6
Nart	13.8	4.0	28.9
Wild populations			
Iraf	12.6	5.6	44.4
Gizel	10.1	5.2	51.5
Dzinaga	11.4	6.4	56.1
Gornaya Saniba	12.2	7.2	59.0
Dargavkij	10.4	5.8	55.8
Synthetic populations			
Syn 305-03	10.8	6.8	62.9
Syn 300-09	11.0	6.0	54.5
Syn 314-08	11.8	5.8	49.1
Syn 316-08	10.9	5.9	54.4
Syn 319-08	12.8	6.4	50.0
Syn 320-08	12.4	6.8	54.8
Syn 321-08	11.6	6.8	58.6
Syn 322-08	12.1	7.2	59.5
TOS-31	11.4	6.5	57.0
SGP-189	12.6	6.0	47.6

Cultivars Daryal (standard), Alan, and Nart decrease the number of stems in the second year of life, which means their poor competitiveness with gramineous and motley grass components. Their adaptation in the mixture is 29–33 percent. Wild forms from highly elevated mountain districts of North Ossetia (villages Dzinaga, Gornaya Saniba, Dargavs) have greater competitive capacity over 50 percent.

Synthetic populations formed from the native species of mountain ecosystems have the maximal competitiveness measured by the number of survived plants.

In the selection of plants for the creation of complex hybrid population, we took into consideration not only competitiveness, but also a group of economically valuable traits: yield of overground mass, high seed productivity, resistance to diseases, fodder merits, longevity, and winter hardiness. Correlations are calculated between all those traits, and it makes possible to select best genotypes in a short period.

Wild samples Dargavs, Iraf, Dzinaga, and Gornaya Saniba, which were a part of synthetic populations, distinguished themselves by good fodder merits. The length of their stems in the phase of stooling was 5–7 cm less than by zoned cultivars. But in the flowering period in mixture crops, the wild forms reached the level of well-known cultivars. They also had advantage in the number of leaves per stem (2–6% above all other studied samples). The rate of leaf cover of stems (58–69%) was the largest among the samples, which originated from the most elevated highlands above sea level. The protein content in plant samples increased under the same rule of vertical zoning of natural habitats.

Biochemical analysis of wild introduced samples revealed that the red clover populations had a high content of protein and a low content of cellulose in the phase of stooling (27 and 14.5%, respectively). In the phase of flowering, the content of protein in the absolutely dry substance reached 19.7–23.2 percent, while the content of cellulose was 17.2–20.1 percent. In the flowering phase, the content of these substances slightly decreased, but it was relatively high in comparison with zoned cultivars. Plants of wild-growing forms contained 0.6–0.8 percent of phosphorus, 2–4 percent of sugar, and 8–10 percent of ashes, which was slightly above the qualitative characteristics of selection samples grown in an elevation of 600 m above sea level (village Mikhailovskoye).

It is important in the selection process of the red clover to create basic material with increased resistance to diseases, especially to root rot, anthracnose, leaf spot of clover, and mildew. With the aim of receiving such cultivar, samples were evaluated in natural conditions of mountains and foothills, on the infectious background, in mixed and pure crops.

Evaluation of samples revealed an advantage of wild forms and complex hybrid populations, which were formed on the base of plants introduced from mountain regions. Synthetic populations Syn 305-03, Syn 321-08, and Syn 322-08 received high estimates for disease resistance (according to the method of All-Russian Institute for Plant Protection).

Incidence of the most widespread diseases in the region (anthracnose, ascochytose, leaf spot of clover) did not exceed 1.5–1.8 points, while other samples were affected on the level of 3.5–4 points.

In the selection of plants in grass mixtures, in the second year of life, the seed productivity was taken into consideration as one of the main traits for cultivars of hayfield-pasture type.

Our studies for many years (1970–2012) revealed that seed productivity varied due to climatic factors. Trials of clover in different agro-ecological zones led us to the conclusion that the optimal period of seed formation is the sum of positive temperatures (above 10°) per vegetation 1,207–1,648° with the quantity of precipitation 445–639 mm, and in the flowering period if seed yield reaches 1.5–1.8 q/ha, the hydro-thermic coefficient (HTC) should be 1.52–3.12. Seed yield decreases to 0.5–1.2 q/ha along with HTC growth.

Unlike steppe and foothill regions, the climate of mountains has its peculiarities. With the ascent to altitudes 1.300–2.200 m above sea level, the short-wave ultraviolet radiation becomes more intense, plant vitality rises, and stimulating influence of ultraviolet rays suppresses partly the negative temperature effect. Under sharp overfall of day and night temperatures active flowering, fruit and seed set are going. Thus, for ripening of seeds in mountain conditions, higher HTC and consequently lesser amount of effective temperatures are needed. It was determined on the results of these studies that with the rise of mountain elevation wild plants (unlike cultivars) had greater percentage of set seeds in comparison with foothills. It is evident that in the mountains low night temperature inhibited transformation of sugars to starch and other substances. It is known that sugar impedes freezing, thanks nectar stores high quality, which is very important for such plants pollinated by insects as clover.

Comparison of native populations, cultivars, and formed complex hybrid populations on different elevation made possible to show the environmental influence on seed set. We took into consideration temperature regime of air and soil, sum of precipitation, humidity and acidity of upper layer of soil (where roots are situated). It was found that on the same elevation, but on different soils, seed set was different. For example, on an elevation of 900 m above sea level, seed set was 27.5 percent on acid soil (pH 4.47) and 46.8 percent on nearly neutral soil (pH 6.45). On an elevation of 2,000 m above sea level on two plots with pH 6.44 and 6.15 seed set of clover was 49.5 and 47.0 percent, respectively. It was also determined that the number of weak seeds changed depending on soil acidity at a maximal proportion (above 50%) at pH 4.47. Seed set in inflorescence is higher on 9.5–27.1 percent, where acidity is 6.0 or above that.

It was found [9] that in selection for seed productivity, it is possible to receive positive results, using phenotypic selection on the following traits: coloration of flowers, number of generative shoots, size of flowering heads, and inflorescences. Seed production depends heavily on the content of starch in the root crown in the period of flowering (correlation coefficient $r = 0.63$), content of sugar in nectar ($r = 0.78$), and presence of pollinators ($r = 0.95$).

1.4 CONCLUSIONS

The complex evaluation method of vertical zoning gives an opportunity of wide phenotypical selection of the forms adjusted to the mountain conditions and creation of new adaptive and productive varieties and restoration of variety of plants on meadows and pastures.

Complex evaluation of selection samples in various conditions growing in mountains and foothills in natural plant communities, in pure and mixed crops secures creation of valuable basic material for formation of a meadow-pasture cultivar with such traits as high competitiveness, qualitative characteristics, and maximal seed productivity.

KEYWORDS

- **Adaptation**
- **Agrocenosis**
- **Clover**
- **Competitiveness**
- **Cultivars**
- **Generative organ**
- **Genotype**
- **Phytocenosis**
- **Populations**
- **Selection**

REFERENCES

1. Shamsutdinov, Z. I.; and Kozlov, N. I.; Importance of genetic collection in intensification of fodder crop selection. *Select Seed-Grow.* **1996,** *3(4),* 9–12. (in Russian).
2. Tyuldyukov, V. A.; Theory and practice of grass farming. Tyuldyukov, V. A.; Moscow: Rosagropromizdat; **1988,** 286 p. (in Russian).
3. Fodder Production in Russia. Moscow: Collection of Works of All-Russian Research Institute for Fodder; **1997,** 428 p. (in Russian).
4. Foster, C. A.; A study of the theoretical expectation of F_1 hybridity resulting from bulk inter-population hybridization in herbage grasses. Foster, C. A.; *Agr. Sci.* **1971,** *76(2),* 293–300.
5. Taylor, N.; Polycrossprogeny tenting of clover (*Trifolium pratense* L.). Taylor, N.; *Crop. Sci.* **1968,** *8(4),* 451–454.
6. Bekuzarova, S. A.; and Dzugayeva, L. A.; A method to define adaptive qualities of red clover's selection samples. Patent 2201076. Published on 20.01.1999 (in Russian).
7. Wide Unified Classificatory of CMEA and International Classificatory of CMEA. Leningrad: Scientific-technical council of CMEA member-states on collections of wild and cultivated species of plants. [Composed by L.V. Leokene et al.] All-Russian Institute for Plant Growing; **1983,** 41 p. (in Russian).
8. Methodic Directions on Selection and Primary Seed-Growing of Perennial Grasses. [Developed by Z.S. Shamsutdinov et al.] Moscow: Publishing House of Russian Agricultural Academy; **1993,** 112 p. (in Russian).
9. Bekuzarova, S. A.; Selection of red clover. Bekuzarova, S. A.; Vladikavkaz: **2006,** 176 p. (in Russian).

CHAPTER 2

INTRODUCED FORMS OF RED CLOVER (*TRIFOLIUM PRATENSE* L.) OF HAYFIELD-PASTURE DIRECTION OF THE USE ON NORTH CAUCASUS

SARRA AB. BEKUZAROVA and VICTOTIA AL. BELYAYEVA

CONTENTS

2.1 INTRODUCTION

Creation of highly productive, genetically valuable forms with high competitiveness in ecosystems of mountain hayfields and pastures is an important element of fodder base formation in the North Caucasian region. Existing in the region, clover cultivars of hayfield-pasture type have low productive longevity, and this fact creates substantial hardships for the formation of agrocenoses in mountain and foothill conditions. The main obstacle in the way of growth of the biological potential of this species is the low adaptive capacity of recommended cultivars in conditions of vertical zoning. The specifics of environmental conditions in mountain regions with billowy relief, where more than half of all agricultural lands are situated in complex topographic conditions characterized by changes of soil-climatic gradients and demands the use of stress-tolerant cultivars. Highly productive cultivars and hybrids are usually less tolerant to such conditions, less effective in conditions of undersow due to low competitiveness with native species, and they have low survival rate of sprouts. Individuals, which have survived, do not live for long and soon fall out the grass. In practice, this leads to unjustified costs of labor and funds [1]. Evaluation and use of genetic potential of local wild populations of red clover has a special importance due to their specific stress tolerance.

Wild mountain species of clover are known for their longevity, frost hardiness, and high content of nutrients. They are more tolerant to high levels of ultraviolet radiation. These species are highly competitive in phytocenosis of mountain hayfields and pastures, and they have a larger number of leaves and shoots per plant. But one should not ignore the rich genetic material of the existing clover cultivars from other regions with wide number of undoubted merits such as high productivity and quality. However, it is necessary to take into consideration that geographic and climatic differences demand the thorough study of introduced forms in conditions of any definite region [2].

Segregation of initial samples for selection is impossible without elaborate biochemical evaluation with the establishment of rate of influence on the parameters under study of the complex of factors affecting plant organism, including stages of development [3]. With the knowledge of amplitude of variation of chemical trait within the limits of population difference, it is possible to choose initial forms for hybridization [1, 4].

In connection with this, the aim of our research consisted in the study of economic and biological traits of introduced forms of the red clover and in selection of samples promising for creation of hayfield-pasture cultivars in North Ossetia—Alania.

2.2 MATERIALS AND METHODOLOGY

We have completed the study and selection of red clover cultivars, introduced from different ecological-geographic natural habitats. Productivity and biochemical content were determined among Belarussian introduced cultivars and samples of red clover (Minskiy Mutant, Ustodlivy, Yaskravy, SL-38, T-46), an introduced cultivar from Siberian region (SibNIIK-10) as well as native wild forms growing in conditions of vertical zoning in North Ossetia—Alania: Dargavski (1,800 m above sea level) and Gornaya Saniba (1,200 m above sea level). Parameters of economic and biological traits of all samples were evaluated in comparison with standard cultivar Daryal, which had been recommended for North Ossetia—Alania from 1993. The authors of the cultivar are S.A. Bekuzarova and B.K. Mamsurov. Biochemical analysis of initial samples of red clover was conducted according to common methods [5]. The following parameters were evaluated: content of dry matter, protein, nitrogen, ash, calcium, phosphorus, cellulose, fat, and sugar. Statistical analysis included descriptive statistics and analysis of variance (ANOVA).

2.3   RESULTS AND DISCUSSION

Biochemical analysis of samples in our study revealed that on the stage of stooling, percent content of dry matter was lower than in the following stages of ontogeny. On the contrary, content of raw protein, potassium, phosphorus, fat, and sugar slightly exceeded similar indices in the stages of budding and beginning of flowering. High content of dry matter and raw protein on the stage of stooling distinguished introduced forms of SL-38 (20.12 and 21.37%, respectively) and Dargavski (19.76 and 20.06%, respectively). The largest content of sugar was found in SL-38 (5.43%) and Gornaya Saniba (4.82%), while calcium and potassium were abundant in standard Daryal (2.08 and 2.87%, respectively). The largest content of ash was observed in SL-38 (12.27%) and standard Daryal (12.66%), phosphorus was abundant in SL-38 (1.09%) and wild introduced forms (0.98–1.02%), and cultivar Yaskravy exceeded all other samples on cellulose content (21.72%) (Figure 2.1, Table 2.1).

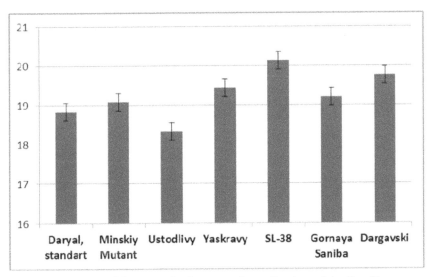

FIGURE 2.1   Content of dry matter in clover samples on the stage of stooling, %.

TABLE 2.1   Chemical content of red clover samples on the stage of stooling, % to dry matter content

Cultivar, sample	Nitrogen	Protein	Ash	Calcium	Potassium	Phosphorus	Fat	Cellulose	Sugar
Daryal (standard)	2.87	17.94	12.66	2.08	2.87	0.86	4.78	17.46	4.55
Minskiy mutant	2.74	17.13	11.84	1.37	2.81	0.84	3.69	17.53	4.34
Ustodlivy	2.25	14.06	11.05	1.41	2.61	0.87	5.71	17.31	4.72
Yaskravy	2.73	17.06	11.23	1.50	2.84	0.91	4.73	21.72	4.16
SL-38	3.42	21.37	12.67	1.53	2.07	1.09	4.85	16.29	5.43
Gornaya Saniba	3.10	19.37	12.13	1.46	2.55	1.02	4.92	16.42	4.82
Dargavski	3.21	20.06	11.24	1.41	2.68	0.98	5.02	15.81	4.69

Results of biochemical content studies of cultural cultivars and native populations of red clover on the stages of budding and beginning of flowering are evident that individual indices have slight positive dynamics in the ontogeny process.

It is known that the energetic component and nutrient value of fodder depends directly on the dry matter content in the yield. On the stage of budding, the largest content of dry matter was observed in samples Minskiy Mutant (21.42%) and SL-38 (21.14%) (Figure 2.2).

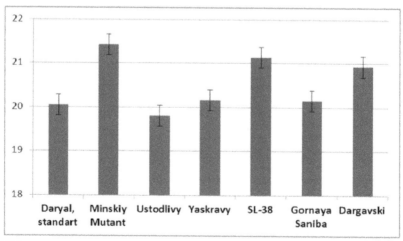

FIGURE 2.2   Content of dry matter in clover samples on the stage of budding, %.

Raw protein is the most valuable component of clover's green mass. It is characterized by good balance of amino acid content. Biological value of clover protein reaches 83–95 percent to the protein of hen's egg. The content of raw protein on the stage of budding was maximal in cultivar SL-38 (16.88%) and minimal in sample Ustodlivy (9.31%) (Table 2.2).

TABLE 2.2   Chemical content of red clover samples on the stage of budding, % to content of dry matter

Cultivar, sample	Nitrogen	Protein	Ash	Calcium	Potas-sium	Phospho-rus	Fat	Cellulose	Sugar
Daryal (standard)	2.33	14.56	10.72	1.72	2.05	0.74	4.66	19.46	2.81
Minskiy mutant	2.10	13.13	10.30	1.29	1.95	0.73	3.21	20.44	3.61
Ustodlivy	1.49	9.31	9.11	1.33	1.62	0.74	5.50	20.31	3.34
Yaskravy	2.31	14.43	10.28	1.38	1.95	0.82	4.55	26.44	3.98
SL-38	2.70	16.88	10.49	1.45	1.04	1.00	4.30	20.65	4.10
Gornaya Saniba	2.43	15.22	10.10	1.40	1.81	0.97	4.71	22.90	3.45
Dargavski	2.37	14.80	9.34	1.38	1.78	0.91	4.80	20.56	3.60

Results of the experiment are evident of the positive correlation in introduced cultivar SL-38 and some other samples between content of dry matter and raw protein.

Besides organic matters, the green mass of clover contains a lot of ash and different elements. In particular, the content of ash in studied samples varied on the stage of budding from 9.11 to 10.72 percent. Standard cultivar Daryal had the best index—10.72 percent.

Potassium plays an important role in the physiological processes of plant organism. It encourages formation of protein substances in plants, increase of sugar synthesis in leaves and their further transportation to other organs, regulates proportion of free and combined water in tissues, and increases hydrophilic effect of protoplasm. Calcium is no less important for the vital functions of plants, and its lack leads to sliming and rot of the root system and loss of tolerance to fungal diseases [6].

According to our data, standard cultivar Daryal contained the largest amount of calcium and potassium—1.72 and 2.05 percent, respectively. Other samples had nearly the same concentration of calcium, but differed on phosphorus accumulation. Cultivar SL-38 contained the largest quantity of phosphorus (1.0%), while among other samples this index varied from 0.73 to 0.93 percent. It is known that phosphorus is a part of nucleoproteins, and reproductive organs of plant and seed embryos are rich in this element. Lack of phosphorus leads to slow growth, depression of flowering, and seed formation. According to our observation, the ratio of calcium and phosphorus varied among different cultivars from 1.4 (SL-38, Gornaya Saniba) to 2.3 (Daryal).

Unlike North Ossetia, in the Leningrad Region, the ratio of these elements average 1.4, which can be considered an optimal level. Such a ratio of calcium and phosphorus is more inherent in another clover species—*Trifolium pratense* L. [7].

Fat is an important elemental part of fodder grasses, and it is necessary to note that its content in legumes is higher than in gramineous plants. Fatty acids in grasses are represented mainly by linoleic and linoleic acids. They cannot be produced in animal organism and should be received with fodder. In our study cultivar Ustodlivy contained a maximal amount of fat (5.50%) at the stage of budding, while sample Minskiy Mutant had a minimal quantity (3.21%).

On the content of cellulose only cultivars Yaskravy (26.44%) and Daryal (19.46%) differed significantly from other samples, among which the amount of cellulose varied in limits 20.31, …, 22.90 percent.

Cultivar SL-38 had the largest content of sugar (4.10%), while standard cultivar Daryal had the smallest content (2.81%).

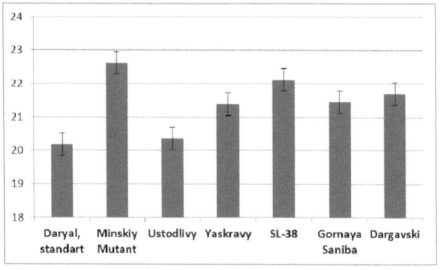

FIGURE 2.3   Content of dry matter in clover samples at the beginning of flowering, %.

The highest content of dry matter at the beginning of the flowering stage was observed in samples Minskiy Mutant and SL-38—22.62 and 22.12 percent, respectively (Figure 2.3).

The content of dry matter in cultivar Yaskravy and samples Gornaya Saniba and Dargavski exceeded by 1.20–1.52 percent the similar index of standard cultivar Daryal, while cultivar Ustodlivy did not differ substantially from standard cultivar. From the stage of budding to the beginning of flowering, the quantity of dry matter grew slightly in all samples of clover.

Insignificant increase of raw protein content at the beginning of flowering in comparison with budding was detected in all samples. Cultivar SL-38 had the highest level of protein (17.88%) and cultivar Ustodlivy had the lowest level (9.75%) (Table 2.3).

TABLE 2.3   Chemical content of red clover samples at the beginning of flowering (% to content of dry matter)

Cultivar, sample	Nitrogen	Protein	Ash	Calcium	Potassium	Phosphorus	Fat	Cellulose	Sugar
Daryal (standard)	2.38	14.89	11.86	1.82	1.92	0.70	4.80	22.54	3.10
Minskiy mutant	2.12	13.25	12.30	1.56	1.80	0.71	3.30	26.68	4.10
Ustodlivy	1.56	9.75	12.26	1.36	1.76	0.68	4.90	28.62	3.80
Yaskravy	2.42	15.13	11.26	1.62	1.67	0.86	4.60	28.56	4.30
SL-38	2.86	17.88	11.12	1.48	1.24	0.95	3.80	22.30	4.50
Gornaya Saniba	2.60	16.22	11.18	1.46	1.90	0.95	4.80	26.77	4.28
Dargavski	2.48	15.53	10.20	1.42	1.81	0.88	4.83	25.94	4.17

Content of ash elements slightly increased in all samples at the beginning of flowering stage. Maximal content of ash (12.3%) was observed in cultivar Minskiy Mutant.

Dynamics of calcium content on the stages of budding and flowering was positive in all clover samples. Maximal content of calcium at the beginning of flowering was observed in cultivar Daryal (1.82%) and the minimal content in cultivar Ustodlivy (1.36%). Dynamics of potassium content was individual in different samples. Potassium content had a positive trend in samples SL-38, Ustodlivy, Gornaya Saniba, and Dargavski and a negative trend in samples Minskiy Mutant, Daryal, and Yaskravy. Cultivar Daryal contained the largest amount of potassium at the beginning of flowering (1.92%).

Phosphorus plays an important role in the formation of reproductive organs of plants. Analysis of its dynamics revealed that its content at the beginning of flowering decreased in comparison with the stage of budding in all samples, except for Yaskravy. Minimal phosphorus content on the stage of flowering was observed in cultivar Ustodlivy—0.68 percent, the maximal content in cultivar SL-38—0.95 percent. Later on at the stage of ripening, we observed further decrease of phosphorus content in the samples.

Clover samples differed substantially in fat content. For example, in cultivar Yaskravy, this index remained at the beginning of flowering practically on the same level as on the stage of budding (change +0.05%). In cultivars Daryal and Minskiy Mutant, fat content slightly increased (+0.11 and +0.09%, respectively). At the same time, fat content substantially reduced in cultivars Ustodlivy (−0.35%) and SL-38 (−0.50%). Despite this reduction, Ustodlivy had the maximal quantity of fat (4.90%).

Quantity of cellulose at the beginning of flowering increased in all clover samples. This index was the highest in cultivar Ustodlivy (28.62%) and the lowest in cultivar SL-38 (22.30%).

It is known that sugar plays an important role in the vital functions of plants, because there is a direct connection between sugar content and processes of amino acid production occurring in conditions of nitrogen assimilation. Besides, the level of frost hardiness depends on the quantity of sugar on late stages of vegetation [8]. It was established in our study that the dynamics of sugar accumulation was positive in all clover samples.

It is possible to state that on the stage of stooling, the green mass of most clover samples has a good quality. On the stages of budding and beginning of flowering (which are especially important for evaluation of economic and biological traits), content of dry matter, protein, ash, calcium, and sugar has an insignificant positive dynamics in all tested samples. At the same time, content of cellulose increased significantly.

Standard cultivar Daryal exceeded all other samples on the stage of budding on content of ash, calcium, and potassium, and at the beginning of flowering on content of calcium and potassium. But it yielded to other cultivars in the content of dry matter, nitrogen, raw protein, phosphorus, fat, cellulose, and sugar.

Belarussian cultivar SL-38 was in the lead on content of dry matter, protein, sugar, and phosphorus. This cultivar along with sample Gornaya Saniba is characterized by an optimal ratio of calcium and phosphorus.

Our studies showed that clover cultivar from Belarus SL-38 and T-46 after introduction to foothill zone displayed high indices of productivity, though they yielded in seed number to wild samples Dargavski and Gornaya Saniba.

A maximal height was observed in standard cultivar Daryal (77.3 cm), the minimal height in SibNIIK-10 (54.0 cm) (Figure 2.4).

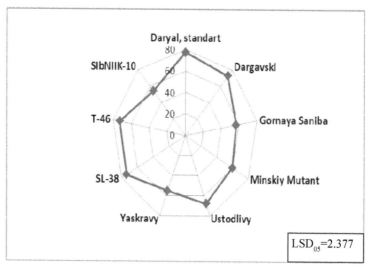

FIGURE 2.4   Height of red clover samples, cm.

Introduced cultivars of Belarussian selection Yaskravy, Minskiy Mutant, and Ustodlivy had medium indices, while T-46 and SL-38 distinguished themselves by greater height −72.7 and 74.0 cm, respectively.

Cultivar SL-38 had the highest productivity of the green mass—3.0 kg/m² (125% to standard), while sample Gornaya Saniba had the minimal productivity—1.1 kg/m² (45.8% to standard) (Figure 2.5). Another native sample Dargavski also had a low productivity of the green mass—1.5 kg/m² (62.5% to standard).

It's necessary to note that usually mountain clovers in their natural habitats have relatively less height. The explanation is that in the conditions of vertical zoning, the existing soil-climatic gradients lead to the formation of agrophytocenoses with maximal adaptation to sharp temperature drops, humidity, sun radiation, and eating by animals.

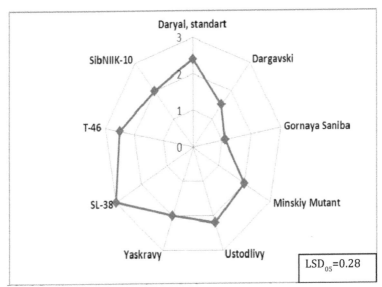

FIGURE 2.5 Yield of green mass in red clover samples, kg/m².

Evidently, natural selection among native clover populations is in the direction of decreasing plant height as an adaptive reaction to more economic consumption of soil nutrients and water resources.

In our study, introduced wild samples of clover did not have low height in all cases. In particular, sample Gornaya Saniba retained this trait in conditions of cultivation on 600 m above sea level. On the contrary, plants of sample Dargavski had larger height than sample Gornaya Saniba. The possible explanation of this difference is the individual type of reaction to meteo-climatic conditions of foothill zone according to adaptive capacities of the sample. Increase of height in sample Dargavski combined with a slight decrease of leaf number per shoot and bushiness. At the same time sample Gornaya Saniba was lower, but had greater bushiness.

Another direction of natural selection among red clover plants in conditions of vertical zoning is the increased seed productivity, which secures effective substitution of perished individuals. We have observed the maximal number of seeds in inflorescence in sample Gornaya Saniba—44 percent (115.8% to standard) (Figure 2.6). A large number of seeds in inflorescences was found also in another wild sample Dargavski—43 percent (113.1% to standard).

Introduced cultivar SibNIIK-10 belongs to cultivars created in ecological-geographic and climatic conditions of Siberian region. Tested for economic and biological traits in conditions of foothill zone of North Ossetia—Alania, it had low height and low seed number in inflorescence, though according to data of cultivars' originators from Siberian Scientific Research Institute for Fodder in its native region the cultivar grew to a height of 83 cm.

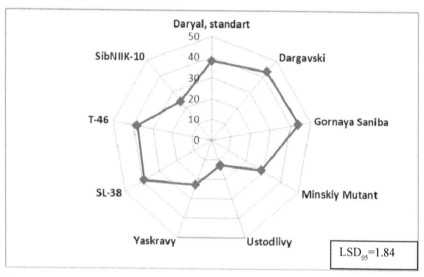

FIGURE 2.6   Seed number in inflorescences of red clover samples, %.

Introduced from Belarus cultivars Yaskravy, Minskiy Mutant, and Ustodlivy had nearly the same productivity, which did not exceed the level of standard. Plants of cultivar Yaskravy had low height and seed number in inflorescences, while in cultivar Ustodlivy the latter index did not exceed 13 percent. Apparently, introduced cultivars SibNIIK-10, Yaskravy, Minskiy Mutant, and Ustodlivy for their economic and biological traits do not suit as initial samples for the creation of hayfield-pasture cultivars in conditions of North Ossetia—Alania, because they lack sufficient adaptation reserves allowing them to adjust to ecological-geographic and climatic conditions of a particular locality.

Our studies showed that standard cultivar Daryal had maximal height. But this cultivar was less productive than SL-38 and T-46 and had less number of seeds in inflorescences than native populations.

Productivity of green mass of wild clover samples Gornaya Saniba and Dargavski is lower in comparison with introduced cultivars, but they have higher percentage of seeds in inflorescences, which is an adaptive reaction to sharp variation of meteo-climatic factors in the highlands.

Introduced cultivar SL-38 displayed itself as the most productive sample. Among the cultivars of Belarusian selection, it has the largest seed number in inflorescences.

2.4 CONCLUSIONS

1. On the stage of stooling, the green mass of most clover samples has a good quality. On the stages of budding and beginning of flowering (which are especially important for the evaluation of economic and biological traits), content of dry matter, protein, ash, calcium, and sugar has an insignificant positive dynamics in all tested samples, while content of cellulose has increased significantly. Dynamics of percentage content of fat, potassium, and calcium varied among samples on different stages of plant development both in positive and in negative direction.

2. The following samples are of greatest genetic value for further selection on economic and biological traits: recommended for North Ossetia cultivar Daryal, introduced from Belarus cultivar SL-38, local wild samples Dargavski and Gornaya Saniba.

3. Cultivar SL-38 has high adaptive capacity. Coincidence of reaction norm of genotype of this cultivar with main ecological and climatic parameters of introduction zone, which is the criterion of successful adaptation, allows its successful introduction in conditions of North Ossetia—Alania and its use in further selection for the creation of hayfield-pasture cultivars along with recommended cultivar Daryal and wild clovers Gornaya Saniba and Dargavski (which have maximal adaptation to highland conditions).

4. Use of mentioned samples in further selection work permits to successfully solve the problem of combination of high yield, seed productivity, nutrient value, and environmental tolerance.

KEYWORDS

- **Budding**
- **Ecological-geographical natural habitats**
- **Hayfield-pasture**
- **Inflorescence**
- **Introduction**
- **Red clover**
- **Seed number**

REFERENCES

1. Bekuzarova, S. A.; Selection of Clover. Bekuzarova, S. A.; Vladikavkaz: Publishing House Gorsky State Agrarian University; **2006,** 176 p. (in Russian).
2. Clover in Russia. Ed. Shpakova, A. S.; Novoselova, A. S.; Kutuzova, A. A.; et al., Voronezh: E. A. Bolhovitinov Publishing House; **2002,** 297 p. (in Russian).
3. Csekat of J., Ribok H. Chrome in of Blakstonia of Koniezyny of Biatei of Poznanskie Copartnership of Sciences. **1989,** *67,* 23–29 p. (in Poland).
4. Bekuzarova, S. A.; Economic biological and biochemical description of standards of world collection of clover grassland. Backlogs of Increase of Production and Use of Forage. Bekuzarova, S. A.; Chebotaeva, S. A.; Ordzhonikidze: Publishing House IR; **1988,** 54–58 (in Russian).
5. Pleshkov, B. P.; Workshop on Plant Biochemistry. Pleshkov, B. P.; Moscow: Publishing House Kolos ("Ear"); **1976,** 250 p. (in Russian).
6. Kul'turnaja Flora: V. XIII. Long-Term Leguminous Herbares. Ed. Muhinoj, N. A.; Stankevich, A. K.; Moscow: Publishing House Kolos ("Ear"). **1993,** 335 p. (in Russian).
7. Baskakova, L. E.; Chemical Composition Clover in Conditions of Leningrad Region. Baskakov, L. E.; Smorodova-Bianchi, G. B.; and Yagunova, G.; Bulletin of All-Union Institute of Plant Growing of N.I. Vavilov. **1975,** *55,* 33–40 (in Russian).
8. Korjakina, V. F.; Physiology of clove and alfalfa. Korjakina, V. F.; Smetannikova, A. I.; Physiology of Clover. Physiology of Agriculture Plants. Moscow: Publishing House of the Moscow State University; **1970,** *4,* 279 p. (in Russian).

CHAPTER 3

RECEIVING OF THE NEW FORMS OF RED CLOVER FOR GROWING IN NORTH OSSETIA, KABARDINO-BALKARIA, AND CHECHEN REPUBLIC

SARRA AB. BEKUZAROVA, IRINA M. HANIYEVA, and LIMA S. GISHKAYEVA

CONTENTS

3.1   INTRODUCTION

Degradation of mountain ecosystems breaches natural processes of energy exchange. It diminishes soil fertility and impoverishes species number of plant organisms. One of the methods of restoration of disturbed landscapes is the undertow of perennial grasses, especially legumes, which have vigorous root system, high content of protein, and capacity to accumulate biological nitrogen in soil [1].

In the intensification of fodder production and making agriculture friendlier to environment, a special role belongs to field foddergrass cultivation. Perennial legumes are an important component of it. Due to their vigorous root system, perennial grasses restrain processes of erosion on slope lands [2, 3].

In modern conditions, creation of seed grass is necessary, especially in mountain conditions on slope lands, where soil and micronutrients are wiped out. Known technologies of legume cultivation include additional fertilization of grass by molybdenum [4]. Molybdenum facilitates increase of nitrogen fixation, development of plant, and in the final result increase of seed yield of clover. A method is proposed to sprinkle the plants by molybdenum–acid ammonium [5]. But several other micronutrients are necessary to increase the activity of tubercles. Besides, molybdenum is sprayed on the stages of budding and flowering, that is, at the time when the grass has already formed, and it decreases the potential of clover plant. Treatment by micronutrients even in small doses on large areas increases costs.

Undersow of legume grasses such as clover, alfalfa, melilot, coronilla, and others is fulfilled for conservation of soil and restoration of fertility on degraded mountain hayfields [3]. This agro-technical process is accomplished on the second way of life, when seed grass is formed. Plants on the second year of life after the harvest of seeds fall out, and as a result biological diversity decreases on mountain hayfields and pastures. Usually two crops—clover (*Trifolium pratense* L.) and alfalfa (*Medicago sativa* L.)—are sown at the same time [5]. These legumes oppress each other in the mixed crop, especially on the second year of life. In the first year after joint sowing those grasses are yet poorly developed. As a result soil erosion takes place. We have used wide-space method of undersow of alfalfa seeds under the grass of clover of the second year of life [6, 7].

Peculiar to legumes hardness of seeds has a significant influence on their germination rate, thus it is necessary to use special methods to increase their germination [8].

Clover plants develop well on slightly acid soils, while alfalfa prefers neutral soils, so for sowing of the second crop alfalfa seeds were mixed with zeolite-containing clay alanite.

Recently, in Russia and abroad, natural zeolites are used successfully to increase the productivity of agriculture and yield of various crops. Such interest in these minerals was evoked by their unique specific property. Carcass aluminosilicates have high resistance to acids and thermal stability. They are also highly active sorbents, cheap selective ionites, and molecular sieves. Their selectivity, molecular-sieve effect, and sorption capacity to cations of alkaline, alkali-earth, rare-earth and several heavy metals, the existence of large deposits, possibility of use before preliminary concentration, and low cost make possible to use zeolites widely in agriculture, industry, and environment protection [9].

In Northern Caucasus zeolite-containing clays (such as irlites, alanites, tereklites, leskenites) are found on flood-plain of river Terek.

3.2   MATERIALS AND METHODOLOGY

On the first year, red clover was sown solidly on the forecrop—winter wheat or winter barley. After the first hay harvest on the second year of life of clover, the wide-space seed grass was formed with the following undersow of alfalfa in row-spacing with lowered norm of sowing. Before sowing alfalfa seeds were mixed with zeolite-containing clay, including alanite (5–6 kg/ha) adding molybdenum-acid ammonium (30–40 g/ha). Hard seeds of alfalfa were treated for 6–8 hr in 0.2–0.3 percent water solution of ragweed. Before sowing wet seeds were enveloped by bacteria from pounded tubercles taken from root system of alfalfa plants of second or third year of life on the stage of budding or flowering. Selection of tubercle bacteria was accomplished on old-age plants in past years. For hectare norm of alfalfa sowing in clover row-spacing (4 kg/ha), just 5–8 plants are needed, in which tubercle bacteria were taken without separation of roots from soil.

3.3   RESULTS AND DISCUSSION

Choice of winter wheat or winter barley as the forecrop for sowing of legume grasses is explained by the fact that those crops leave in soil unused microelements (boron, molybdenum, cobalt, copper, manganese), which promote the increase of chlorophyll content in leaves, strengthening of plant's assimilation activity, and process of photosynthesis. They also influence positively on seed development and their sowing qualities. Under influence of micro-elements, plants become more resistant to various diseases and unfavorable environmental conditions (soil and aerial drought, excessive moisture, high and low temperatures). For legume grasses, clover, and alfalfa, substantial quantity of microelements is needed. Placement of legume grasses after winter cereals permits to cut costs on micronutrients and increase their fodder and seed productivity [5].

From the beginning, we have sown clover solidly to cut the number of weeds in the crop. The next year on the stage of budding, we received fodder of good quality in the hay harvest. After the first hay harvest, we formed from solid sowing of clover with the help of cultivator a wide-space sowing with row-spacing 60 cm wide, where alfalfa was undersown with lowered norm of sowing 4 kg/ha. This norm of sowing secures better lighting of grass, increase of quantity of generative organs, and more active pollination by bees. Additional fodder is received in the year of alfalfa sowing, in the end of its vegetation. The following year (the third year of clover's life), the first legume crop (clover) falls out, and seed grass of alfalfa remains. It is harvested in the first hay-crop.

Due to the fact that clover plants develop well on slightly acid soils (pH 5–6), while alfalfa grows better on neutral soils (pH 7–7.5), special technique was applied for sowing the second crop. Seeds of alfalfa were mixed with zeolite-containing clay alanite, where molybdenum–acid ammonium was dissolved preliminarily. High content of calcium in alanite (30–40%) leads to decrease of soil acidity, and it has a favorable influence on alfalfa development.

Alfalfa grass develops a vigorous root system, securing yields of fodder and seeds for 6–8 years with simultaneous increase of soil fertility due to legume grasses' capacity to accumulate biological nitrogen.

Mixture of zeolite-containing clay alanite with molybdenum–acid ammonium leads to normalization of soil solution and decreases acidity in seedbed. Thus, joint treatment by molybdenum–acid ammonium and alanite is necessary for the development of alfalfa plants. Zeolite-containing clay prolongs the action of the mixture. It slowly releases elements contained in the mixture.

The above-mentioned agro-technical process—treatment of seeds before sowing in the water solution of ragweed and enveloping of wet seeds by tubercle bacteria from the root system of alfalfa—has significant importance. This process permits to increase germination rate of seeds and energy of sprouting as well as to decrease seeds' hardness.

Ragweed (*Ambrosia artemisiifolia* L.) contains the complex of chemical substances, including essential oil, glycosides, macro-elements, which stimulate seed germination. Unlike other phyto-stimulators (nettle, wormwood, etc.), ragweed grows everywhere and is often found as a weed in various crops.

Results of the experiment revealed that germination rate of alfalfa seeds grows from 72 percent (control—soaking in the water) to 92 percent in the optimal way of ragweed treatment. At the same time, hardness of seeds decreased from 13 to 8 percent in the optimal variant. As a result, erosion processes on slope lands weaken 2–3 times.

3.4   CONCLUSIONS

1. Placement of legume grasses after winter cereals secures increase of seed yield of legume grasses.
2. Treatment of seeds before sowing by mixture of zeolite-containing clay and molybdenum–acid ammonium by ragweed solution facilitates decrease of hardness of legume grasses' seeds.
3. Additional fertilizing of seed grass by water solution of ragweed increase seed yield of clover and alfalfa on 18–25 percent.
4. Using natural resources (zeolite-containing clay, weed plant ragweed), it is possible without additional costs to increase seed yield on slope lands and weaken erosion processes 2–3 times.

KEYWORDS

- **Alanite**
- **Alfalfa**
- **Ceolit**
- **Clover**
- **Coronilla**
- **Foddergrass**

- **Irlit**
- **Legume grasses**
- **Leskenit**
- **Molybdenum**
- **Tereklit**

REFERENCES

1. Gazdanov, A. U.; Bekuzarova, S. A.; and Yefimova, V. A.; Destructive Processes on Mountain Hayfields and Methods of Their Melioration. Vladikavkaz: Publication of Gorsky State Agrarian University; **2006,** 96 p. (in Russian).
2. Vasin, V. G.; Vasin, A. V.; and Kiselyova, L. V.; Perennial grasses in pure and mixed sowing in the system of green conveyor. *Kormoproizvodstvo (Fodder Production).* **2009,** *2,* 14–16 (in Russian).
3. Bekuzarova, S. A.; and Gasiyev, V. I.; Fodder Crops in North Ossetia—Alania. A Monograph. Vladikavkaz: Publication of Institute of Raising the Levels of Instructors' Skills 'Mavr'; **2012,** 148 p. (in Russian).
4. Novosyolova, A. S.; Selection and Seed-Growing of Clover. Moscow: Agropromizdat; **1986,** 174–175 (in Russian).
5. Bekuzarova, S. A.; Aznaurova, Zh. U.; and Tsagarayeva, E. A.; The method of non-root treatment of seed grass of clover. Patent for invention # 2189719 published on September 27, **2002.** Bulletin # 27 (in Russian).
6. Bekuzarova, S. A.; Bekmurzov, A. D.; Basiyev, S. S.; Shogenov, M. L.; and Pliyev, Y. V.; The method of sowing of legume grasses in crop rotation. Patent for invention # 2155463 published on September 10, **2000.** Bulletin # 25 (in Russian).
7. Adayev, N. L.; Bekuzarova, S. A.; and Abasov, S. M.; The method of placement of legume seed grass on slope lands. Patent for invention # 2425476 published on August 10, **2011** (in Russian).
8. Zherukov, B. H.; Khaniyeva, I. M.; Khaniyev, M. H.; Magomedov, K. G.; Bekuzarova, S. A.; and Boziyev, A. L.; The method of alfalfa seed treatment before sowing. Patent for invention # 2479974 published on November 24, **2013** (in Russian).
9. Postnikov, A. V.; Loboda, B. P.; and Sokolov, A. V.; Use of zeolites in agriculture. *Bull. Russ. Acad. Agric. Sci.* **1992,** *5,* 49–51 (in Russian).

CHAPTER 4

PRODUCTIVITY OF AMARANTH OF CULTIVAR IRISTON IN THE CONDITIONS OF NORTH OSSETIA—ALANIA

SARRA AB. BEKUZAROVA and VADIM IR. GASIEV

CONTENTS

4.1   INTRODUCTION

Finding ways to improve the use and introduction of nontraditional plants to create new food products, as well as facilities for the pharmaceutical and cosmetic industry, is an urgent requirement. One of these objectives in the mountains and foothills of the Republic of North Ossetia—Alania is the culture of amaranth, in particular, creating a new cultivar of Iriston [1, 2].

The productivity of forage crops, including amaranth, primarily depends on the individual plant stand of grass, which is highly dependent on the method of planting and seeding rate [3]. Creating optimal density grass is the key to obtain high yield. In the initial phases of development amaranth plants grow slowly. In this regard, special attention should be paid to the treatment of soil, retaining moisture in the top layer, because the depth of seed placement is small (1–2 cm). Germination begins with the formation of plant density, which in turn allows estimation of field germination and preservation of plants [3–5].

4.2   MATERIALS AND METHODOLOGY

For the purpose of increasing the efficiency of a new grade of an amaranth of Iriston, we studied ways of crops at row-spacings of 15 cm (narrow-rowed) and 45 cm (wide-row) with norm of seeding of seeds 1.5, 1.75, and 2 million pieces/ha. At this we studied viability of seeds, proportion of surviving seeds (%), density of grass stand, photosynthetic activity, considering the area of leaves (AL), and the photosynthetic potential, productivity of green material and seeds from unit of area. The photosynthetic potential is an indicator, which is used for an assessment of a condition of crops [5]. It represents the sum of daily indicators of the area of leaves for a certain period on unit of area. The area of each allotment made 30 m² in quadruple frequency. Green material was analyzed according to the content of nutrients (fodder pieces, a digestible protein, solid, exchange energy).

4.3   RESULTS AND DISCUSSION

As a result, it was shown that the viability of seeds and plant survival completely dependent on seeding rate and method of sowing culture (Table 4.1).

TABLE 4.1   Influence of ways and norms of seeding on field viability and safety of plants of an amaranth of a grade of Iriston (an average for 2010–2012)

Way of crops	Norm of seeding (one million pieces/ha)	Density of plants (one thousand pieces/ha)		Field germination (%)	Safety (%)
		In a phase of shoots	Before harvesting		
15 cm (ordinary way)	1.5	944.4	220.6	63.0	23.4
	1.75	1107.6	242.3	63.3	21.9
	2.0	1270.8	267.5	63.5	21.1
45 cm (wide row)	0.5	318.2	135.6	63.6	42.6
	0.75	482.1	156.8	64.3	32.5
	1.0	663.5	174.4	66.4	26.3

In our research, viability depends on the density of the sowing, which in turn depends on the norm of sowing. The higher the norm of sowing, the denser sowing will be both in the ordinary method of sowing and in wide-row. On the average, for 3 years of researches, the lowest field viability was observed in an ordinary method of sowing crops, and at increase in norm of seeding it raised from 63.0 to 63.5 percent. In wide-row crops with wash seeding of seeds to 1.0 million pieces/ha, crops were characterized by the maximum viability—66.4 percent. An increase in norm of seeding mutual oppressing action, which leads to considerable, thin crops and decrease in safety of plants of an amaranth, is observed. Supervision over plants of the first year of researches to the third allowed establishing optimum terms of crops of an amaranth in the conditions of the Republic of North Ossetia—Alania (Table 4.2).

TABLE 4.2 Influence of ways and norms of seeding on field viability and safety of plants of an amaranth of a grade of Iriston (2010–2012 years)

Way of crops	Norm of seeding (one million pieces/ha)	Density of plants (one thousand pieces/ha)		Field germination (%)	Safety (%)
		In a phase of shoots	Before harvesting		
2010 (first year)					
15 cm (ordinary way)	1.5	963.0	229.8	64.2	23.9
	1.75	1129.0	250.6	64.5	22.2
	2.0	1293.5	271.3	64.7	21.0
45 cm (wide row)	0.5	320.3	135.2	64.1	42.2
	0.75	491.7	157.9	65.6	32.1
	1.0	673.2	179.3	67.3	26.6
2011 (second year)					
15 cm (ordinary way)	1.5	927.8	212.0	61.9	22.8
	1.75	109.3	234.2	62.4	21.5
	2.0	125.2	262.7	62.8	20.9
45 cm (wide row)	0.5	316.9	134.8	63.2	42.7
	0.75	472.9	154.3	63.1	32.6
	1.0	651.7	173.2	65.2	26.6
2012 (third year)					
15 cm (ordinary way)	1.5	942.3	220.1	62.8	23.4
	1.75	1102.5	241.3	63.0	21.1
	2.0	1263.7	268.4	63.2	31.2
45 cm (wide row)	0.5	318 .4	136.7	63.7	42.6
	0.75	481.7	158.3	64.9	32.9
	1.0	665.6	170.2	66.6	26.3

Thus, an increase of norm of seeding in ordinary crops on the average for 3 years of researches, the safety of plants of an amaranth decreased from 23.4 to 21.1 percent, and on wide-row from 42.6 to 26.3 percent.

Efficiency of plants completely depends on photosynthetic activity. From intensity of photosynthesis in direct dependence, there is as a whole a development of plants, which is closely connected with the size and period of operation of the sheet device [3, 4].

Researches of dynamics of formation of a sheet surface of plants of an amaranth showed that density of crops had essential impact on its assimilatory square and duration of functioning. With increase of norm of seeding of seeds, the difference increased in the size of an assimilatory surface. First, the closely connected process of photosynthesis with the formation of a biomass of a plant is explained. Despite this, in initial growth phases and developments of plants of an amaranth, there is a slow increase of its sheet weight that defines weak competitiveness in relation to weed vegetation (Table 4.3).

TABLE 4.3   Photosynthetic activity agrocenoosis an amaranth depending on norms and ways of crops

Way of crops	Norm of seeding (one million pieces/ha)	Year of research						Average	
		2010		2011		2012			
		AL	PP	AL	PP	AL	PP	AL	PP
15 cm (ordinary way)	1.5	36.0	0. 67	50.6	0.82	40.8	0.72	42.5	0.74
	1.75	36.9	0.69	51.2	0.86	41.2	0.73	43.1	0.76
	2.0	37.7	0.70	52.3	0.90	42.0	0.76	44.0	0.79
45 cm (wide row)	0.5	37.4	0.72	50.4	0.83	39.9	0.77	42.6	0.77
	0.75	38.6	075	51.2	0. 86	40.7	0.82	43.5	0.81
	1.0	39.7	0.78	52.5	0.92	42.3	0.87	44.8	0. 86

The maximum leaf area and photosynthetic potential fall on the phase of maturing of seeds. In ordinary crops, the largest area of leaves is formed by crops with norm of seeding of 2 million pieces/ha, which were, respectively, on the average for years of researches 44.0 m^2/ha, in wide-row crops at norm of seeding of 1 million pieces/ha the area of leaves were 44.8 m^2/ha.

Thus, the increase in norm of seeding, conducts to consolidation of crops. In such agrocenosis, photosynthetic potential was much higher in comparison with thin crops that were caused by a high sheet surface of plants. However, in ordinary crops, depending on growth phases and development of an amaranth, the tendency to decrease the photosynthetic potential in condensed agrocenosis was noted.

Photosynthetic potential was the highest in ordinary crops, at the maximum norm of seeding of seeds in a phase of full maturing, which averaged 0.79 million m^2 of days/ha in three years of researches. In wide-row sowing (45 cm), the photosynthetic potential of plants varies between 0.77-0.86 million m2 days/ha.

Forage mass was dependent on the methods of planting and seeding rates in the foothills of the Republic of North Ossetia—Alania (Table 4.4).

TABLE 4.4   Productivity of green material of an amaranth of a grade of Iriston, tons/ha (2010–2012)

Way of crops	Norm of seeding (one million pieces/ha)	Year of research			Average for 3 yr
		2010	2011	2012	
15 cm (o r d i n a r y way)	1.5	43.3	38.7	37.2	40.0
	1.75	46.0	40.4	40.2	42.2
	2.0	48.9	43.5	42.9	45.1
45 cm (wide row)	0.5	43.8	39.1	38.6	40.5
	0.75	46.4	42.0	42.3	43.6
	1.0	49.7	44.8	43.9	46.1
SED$_{05 \text{ tons/ha}}$	-	1.6	1.3	1.9	–

Important indicator of fodder culture of an amaranth is its quality indicators presented in Table 4.5.

TABLE 4.5   Quality indicators of an amaranth (on the average for 2010–2012)

Way of crops	Norm of seeding (one million pieces/ha)	Solids (tons/ha)	Fodder pieces (tons/ha)	Digestible protein (tons/ha)	Exchange energy (GJ/ha)
15 cm (ordinary way)	1.5	6.9	5.6	1.08	72.70
	1.75	7.4	6.1	1.16	77.32
	2.0	7.9	6.5	1.24	82.95
45 cm (wide row)	0.5	7.3	5.9	1.10	80.62
	0.75	7.9	6.3	1.18	85.83
	1.0	8.5	6.7	1.28	89.96

By researches, it is established that the highest productivity of green material of an amaranth private (15 cm) created crops with norm of seeding of 2 million pieces/ha and wide-row (45 cm) with norm of 1 million pieces/ha on the average for years of researches, respectively, 45.1 and 46.1 tons/ha.

The greatest productivity of ordinary crops of an amaranth is provided, first, at the expense of optimum distribution of seeds on unit of area, thanks to the optimum density of herbage created.

The increase in productivity of wide-row crops of an amaranth allows cutting twice expenses on a sowing material.

On the yield of biomass, amaranth depend on the yield of dry matter, of fodder pieces of digestible protein and energy. It is established that in wide-row planting methods, the yield of dry matter increases from 7.3 up to 8.5 tons/ha and in the ordinary crops of this indicator over the years, studies amounted to 6.9 and 7.9 tons/ha. A similar trend was observed in the output of fodder pieces; in ordinary (15 cm) crops, this indicator ranged from 5.6 to 6.5 tons/ha in wide-row sowings of 5.9–6.7 tons/ha, digestible protein 1.08–1.24 tons/ha, in the ordinary crops. Wide-row is 1.10–1.28 tons/ha. In the ordinary crops with the seeding rate of 1.5–2.0 million pieces/ha, output energy amounted to 72.7–82.95 GJ/ha, and under wide-row planting method with the number of seedings from 0.5 to 1.0 million pieces/ha—80.62–89.96 GJ/ha.

Also, it was found that the yield of seeds of amaranth depends on the method of planting and seeding rate. In ordinary (15 cm) and wide-row (45 cm) crops, increased seeding rate leads to a decrease in productivity of seeds. Qualitative indicators of forage depend on the climatic conditions of the year change (Table 4.6).

TABLE 4.6   Nutritious forage mass of amaranth (2010–2012)

Way of crops	Norm of seeding (one million pieces/ha)	Solid (tons/ha)	Fodder pieces (tons/ha)	Digestible protein (tons/ha)	Exchange energy (GJ/ha)
2010 (first year)					
15 cm	1.5	7.8	6.4	1.16	79.83
(ordinary way)	1.75	8.3	6.8	1.26	85.78
	2.0	8.9	7.3	1.33	90.53
45 cm	0.5	8.0	6.7	1.22	81.19
(wide row)	0.75	8.5	7.1	1.29	89.67
	1.0	9.2	7.7	1.37	92.30
2011 (second year)					
15 cm	1.5	6.6	5.3	1.03	69.54
(ordinary way)	1.75	7.0	5.8	1.12	74.03
	2.0	7.5	6.1	1.20	79.96

TABLE 1   *(Continued)*

Way of crops	Norm of seeding (one million pieces/ha)	Solid (tons/ha)	Fodder pieces (tons/ha)	Digestible protein (tons/ha)	Exchange energy (GJ/ha)
45 cm (wide row)	0.5	7.1	5.4	1.03	80.72
	0.75	7.8	5.9	1.13	84.69
	1.0	8.3	6.3	1.24	89.31
2012 (third year)					
15 cm (ordinary way)	1.5	6.4	5.2	1.02	68.73
	1.75	6.8	5.6	1.09	72.14
	2.0	7.3	6.0	1.19	78. 37
45 cm (wide row)	0.5	6.9	5.5	1.04	79.94
	0.75	7.5	5.8	1.11	83.13
	1.0	8.1	6.2	1.22	8826

The greatest yields of seeds provided the option wide-row sowing method (45 cm) with norm of seeding of 0.5 million pieces/ha, which averaged over the years by 1.34 tons/ha, whereas the average crop yield decreased by 0.78–0.9 tons/ha.

In the course of analysis of seed productivity of amaranth, it was revealed that in ordinary crops, seed yield was less in comparison with wide-row crops (Table 4.7).

TABLE 4.7   Seed productivity of amaranth depending on the methods of sowing and seed rate, tons/ha

Way of crops	Norm of seeding (one million pieces/ha)	Year of researches			
		2010	2011	2012	Average
15 cm (ordinary way)	1.5	0.49	0.58	0.60	0.56
	1.75	0.41	0.49	0.53	0.48
	2.0	0.37	0.47	0.48	0.44
45 cm (wide row)	0.5	1.19	1.44	1.38	1.34
	0.75	1.12	1.33	1.29	1.25
	1.0	1.02	1.30	1.20	1.14
SED $_{05tons/ha}$	-	0.09	0.56	0.23	-

During all years of studies on wide-row sowings, plants were formed with more height and weight than the member planting method. Thus, the weight of plants in wide-row sowings were above ordinary level at 83–91 cm, plant height superiority of crops with aisle 45 cm above the ordinary (45 cm) is 18–19 cm, respectively. The observed changeability in the different years of research is summarized in Table 4.8.

TABLE 4.8   Average weight of one plant amaranth, g (2010–2012)

Way of crops	Norm of seeding (one million pieces/ha)	2010 г	2011 г	2012 г	Average
15 cm (ordinary way)	1.5	112	103	106	107
	1.75	108	100	102	103
	2.0	103	95	96	98
45 cm (wide row)	0.5	196	201	198	198
	0.75	188	193	190	190
	1.0	172	185	186	181

The results of 3 years of studies revealed that the highest values of the average mass of one plant amaranth was obtained in option wide-row seeding with norm of seeding of 0.5 million pieces/ha, which gives the basis to recognize the specified method of planting and sowing rate promising for the formation of high yields in the conditions of North Ossetia—Alania.

4.4   CONCLUSION

The optimal way of seeding amaranth cultivar Iriston in the Republic of North Ossetia—Alania is wide-row sowing with inter-rows 45 cm with norm of seeding of 0.5 million pieces/ha. Such parameters increase yield of green mass and quality and weight of seeds of valuable fodder culture.

KEYWORDS

- **Agrocenosis**
- **Amaranth**
- **Forage**
- **Photosynthetic potential**
- **Yield**

REFERENCES

1. Bekuzarova, S. A.; and Gasiev, V. I.; Fodder Crops in North Ossetia-Alania. Vladikavkaz: Moor; 2012, 147 p. (in Russian).
2. Chernov, I. A.; Amaranth—perspective source of fodder protein. *Bull. Agric. Sci.* **1992**, *2*, 82–86. (in Russian).
3. Zelenkov, V. N.; Gylshina, V. A.; and Tereshkina, L. B.; Amaranth. Agro-Portrait. Moscow: Ros. Academy Natural Sciences; 2008, 103 p. (in Russian).
4. Belikova, S. V.; Haevskaya, P. P.; and Podkolzin, A. I.; Experience of Cultivation of Amaranth in the Stavropol Region Materials of the 1st All-Union Scientific Conference. The Cultivation and Use of Amaranth in the USSR. Kazan: Publishing House of Kazan state University; 1991, 37–46. (in Russian).
5. Gromov, A. A.; Amaranth in the South Urals Grassland. **1995**, *4*, 28–32. (in Russian).

CHAPTER 5

VARIETAL POTATOES FEATURE DURING DIFFERENT TIMES OF PLANTING IN FOOTHILL ZONE OF NORTH OSSETIA ALANIA

SOLTAN S. BASSIEV

CONTENTS

5.1 INTRODUCTION

Many factors forming the net result depend on time of potatoe planting: productivity and tuber quality. Therefore, when determining this time, it is necessary to consider weather, soil condition, physiological, and condition of tubers [1, 2]. The time of planting influences plant treatment in summer. Each cultivar needs to be planted in one place within the shortest time (up to 7 days); otherwise, the subsequent treatments with pesticides may be inefficient. Not only too late planting of potatoes is harmful, but also at very early plating there is a threat of frosts to crops. At the same time, at early time of plating, before activating the lice carrier, the plant reaches age stability, and therefore, suffers less from virus diseases [1–3].

Due to these arguments, we in this chapter have studied the phenology of potato plant development, accumulation of a crop, and quality formation depending on the time of planting.

5.2 MATERIALS AND METHODOLOGY

Researches and observations were made on the "State Farm by Dzerzhinsky" in Alagirsky region of RNO-Alania on pebbling leached chernozem. Analyses of indicators for soil saturation with nutrients and humus on the experimental plot revealed that soils contain quite high percentage of humus, 52–63, and contain phosphorus, 0.44–0.21, and potassium, 1.62–2.04 mg equivalent per 1 kg of the soil. Sowing was performed for three data: 15.04, 25.04, and 05.05.

Cultivars of local selection Vladikavkazsky and Predgorny (North Caucasian Research Institute of Mountain and Foothill Agriculture) were used for researches; cultivar Volzhanin was applied as a cultivar-standard.

We followed the subsequent conditions of landing of potato: agrotechnology was universally adopted, depth of potatoes planting was 6–8 cm, the total area of the plot was 28 m^2, and that of record plot was 25 m^2, the frequency of replication was quadruple, and the preceding crop was winter wheat.

For planting we used tubers weighing 60–80 g. The nitric–phosphoric–potassium fertilizer (in Russian—"nitroamophoska") (in correlation 32:32:32) was introduced in spring. Care for plantings was carried out according to the time of planting and also depending on weed and soil condition. Against buckeye rot and Colorado beetle, we applied double treatments with preparations Karate and Ridomil-Gold, respectively.

The following observations were carried out during the course of the researches:
- phenological observations as calculation of the main stalks in a bush; observations for growth of a top and its test in a phase of potatoes flowering;
- counted assimilatory surface of leaves by cutting method;
- determined the content of starch and dry matters by weight method;
- determined tubers' marketability.

Crop accounting was carried out from each plot, and we received results processed by the method of dispersion analysis [4].

5.3   RESULTS AND DISCUSSIONS

As our researches show (Table 5.1), the time of emergence of seedlings and further approach of growth phases and development of potato plants depend on the biological features of cultivars, the agrotechnical methods, and the weather conditions that developed during the vegetation period.

TABLE 5.1   Phenological observations for various cultivars during different time of planting

Time of planting	Volzhanin			Vladikavkazsky			Predgorny		
	Sprouts	Budding	Flowering	Sprouts	Budding	Flowering	Sprouts	Budding	Flowering
1	20	38	43	23	42	46	19	36	40
2	17	36	40	20	40	43	16	34	37
3	14	33	37	17	37	39	13	31	33

First, seedlings appeared depending on the early ripening of a cultivar.

According to all planting times, we noted that the emergence of seedlings of cultivar predgorny occurred 2–3 days earlier than cultivar Volzhanin and 3–6 days earlier than cultivar Vladikavkazsky.

It is revealed that when the later planting was performed, the higher was the temperature of the soil; thus, the emergence of both seedlings and other phases of development of potato plants began earlier.

The bush of potato plant consists of several rather autonomous stalks, whose quantity is a varietal sign. As a rule, there is a positive dependence between the number of stalks and quantity of the formed tubers; however, a direct link is quite often absent.

Our studies revealed that potatoes cultivars had unequal quantity of the main stalks in a bush. According to Table 5.2, this sign depends on the cultivar biology. Special changes from time of planting have not been noted—0.3–0.6 piece/bush in every cultivar.

TABLE 5.2   Number of stalks and height of plants in dependence on the cultivar and time of planting in phase of flowering

Date of planting	Volzhanin		Vladikavkazsky		Predgorny	
	Number of stalks	Height of plants (cm)	Number of stalks	Height of plants (cm)	Number of stalks	Height of plants (cm)
1	3.4	57	4.0	59	4.0	60
2	3.6	53	4.2	57	3.9	59
3	3.5	50	4.0	51	3.7	53

Note: 1—15.04; 2—25.04; 3—05.05

Measuring the height of potato plants revealed that cultivar Predgorny is distinguished by taller plants, cultivar Volzhanin slightly lesser and finally cultivar Vladikavkazsky (i.e., this indicator is not connected with the group of ripeness and is especially an individual sign depending only on genetic features of this cultivar).

Plants height depending on cultivation conditions was subjected to small fluctuations. First, it is possible to note the influence of weather conditions during the vegetative period. The difference in plant height depending on the cultivar and time of planting is insignificant, but has some regularity. In the first time of planting, height of potato bushes in all cultivars was 1–5 cm more than the second variant and 7–8 cm than the third. In our opinion, these distinctions are connected with moderate air temperature during vegetation of 1 and 2 times of planting. Besides, favorable conditions for growth and development of potato plants were created; atmospheric precipitation was more densely used.

Well-developed healthy top provides formation of high tuber yield. Certain relation exists between the maximum mass of a bush leaves and the yield. Many authors attach great value to the size of a leaf surface by the beginning of tuber formation, considering that the more it is during this period, the larger tuber yield will result [5, 6].

There is no consensus with reference to the influence of top weight on the yield of potato tubers in scientific literature. A number of researchers note that equally with increase in weight of the top productivity of tubers also increases [1, 6]. Other authors state that the final tuber yield not always depends on the power of top development [2, 3, 5]. The numerous experiments conducted in various soil and climatic zones of the country with cultivars of various ripening groups allowed A. G. Lorkh [6] to conclude on expediency of the directed top development (by introduction of fertilizers in proper doses and watering) for receiving maximum tuber gain.

In our experience, when planting potatoes in different times, growth, development, and formation of an overground biomass took place under unequal environmental conditions, which was considerably reflected in the duration of top functioning and work efficiency.

During the study period, in the initial period, when weather conditions favored the growth and development of overground parts of potato plants, the top developed evenly according to the time of planting.

The largest number of top was collected during the optimum time on the second variant, and the mass of the top in a phase of flowering for the cultivar Volzhanin was 440, 580, and 470 g/bush, whereas in early and late times of planting 100–120 g/a bush less.

If we consider the accumulation of top mass in view of cultivars, it is possible to note that the most powerful top was formed with the cultivar Vladikavkazsky. The least top mass during vegetation in the period of researches appeared at the cultivar Volzhanin; however, it conceded not much to the cultivar Predgorny.

As potato is a photophilous plant, with insufficient illumination it forms long and thin stalks: the top turns yellow, the flowering is late, and the efficiency of photosynthesis decreases. All this leads to a decrease in potatoe yield and deterioration of its quality. Therefore, during the cultivation of potatoes, it is important to create conditions of normal illumination for each plant.

Results of our researches (Table 5.3) revealed that in the phase of full plant flowering, the most optimum area of the leaf surface occurred in cultivar Vladikavkazsky (34.3,000 m²/ha) on the second time of planting. At the same variant at cultivars Volzhanin and Predgorny, the assimilating surface was 32.1 and 33.4,000 m²/ha, respectively.

TABLE 5.3 Top mass and assimilating surface of various potato cultivars depending on the time of planting

Time of planting	Volzhanin		Vladikavkazsky		Predgorny	
	Mass (g/bush)	Assimilating surface (thousand m²/ha)	Mass (g/bush)	Assimilating surface (thousand m²/ha)	Mass (g/bush)	As-similating surface (thousand m²/ha)
1	370	30,1	490	29,9	390	31,3
2	440	32,1	580	34,3	470	33,4
3	360	28,3	470	27,4	350	29,7

In agricultural literature, a lot of contradictory statements concerning the efficiency of planting time per potato tuber yield have been recorded. This is probably because the climatic conditions and the cultivar role in determination of their influence on productivity are not always considered. On soils with various mechanical compositions, spring high-plastic condition and its warming up do not occur at the same time—from 5 to 20 days. In areas with prevalence of light sandy and sandy loam soils, the high-plastic condition comes earlier than the accepted recommendations [3].

As a rule, yield loss because of the delay of potatoes planting to 20 and more days is not compensated with the increase in planting density and additional application of fertilizers [2, 3]. One of the important arguments against early potato planting is the fear of sprouts getting damaged because of late spring frosts. However, in the larger territory of the Central Nonchernozem Region, plants seldom come under frosts at early planting, but decrease in yields because retarded planting is observed annually [6].

Potato planting in different calendar days practically does not influence expenses, but productivity rate from return at early plantings promotes increase of relative net profit of 44–49 percent [1, 6].

High efficiency of crops can be received at the optimum combination of all factors influencing the formation of productivity. Therefore, agrotechnical actions have to be strictly specified for certain soil-climatic conditions, and cultivated cultivars as well.

In our experience, more stable and rather high yield of tubers in all the three studied cultivars have been noted.

Various cultivars reacted differently to the developed weather conditions, clearly owing to the biological features. The highest yields in all the three cultivars were recorded on the second planting. In our experience, the cultivar Predgorny appeared to be the most high-yielding plant (25.8 t/ha) and Vladikavkazsky (24.3 t/ha) slightly less. On the first planting time, yields were 2.2–3.1 t/ha less, than on the second.

Various tuber marketability is noted on cultivars. The lowest is on the cultivar Volzhanin, fluctuations of which depend on planting times made 4–8 percent. By optimum variant its marketability reached 82 percent.

Researches showed that the cultivar Predgorny had higher marketability than other cultivars, thus maximum was received on the second planting time, 96 percent. In other cultivars, this indicator did not reach even the lower limit of the cultivar Predgorny.

One of the main quality indicators of potato tubers is starch and dry matters, which are determined by a number of factors: varietal features, climatic conditions, agrotechnology, rate, form and relation of fertilizers, degree of tubers maturity, etc.

A certain correctional connection between the content of dry matters and starch in the tuber is determined. Depending on the varietal features and early ripening of the cultivar, the difference between the content of dry matters and starch reaches 4.6–11.0 percent. Deviations depending on the cultivar of one ripening group reached 6.0–7.0 percent [7].

S. T. Prokoshev [8] determined that mid-ripening and late-ripening potato cultivars always have higher content of starch than early-ripening cultivars, irrespective of the applied agrotechnology.

Contradictory opinions about influence of planting time on tuber starchiness are encountered in the literature. Thus, the majority of authors state that the maximum starch accumulation is in variants with early planting [5–8].

In the process of plant growth and development, accumulation of dry matters is unstable as a result of fluctuations of weather and other conditions.

According to Table 5.3, the content of dry matter and starch in tubers do not significantly differ according to the variants or the cultivars. For all cultivars, optimum according to these factors was the early planting.

The largest amount of dry matters and starch were accumulated by the cultivar Predgorny, 19.3 and 14.2 percent, respectively. The cultivar Volzhanin during the third planting accumulated less, 15.4 and 11.3 percent of dry matters and starch, respectively.

The cultivar Vladikavkazsky accumulated dry matters and starch 0.5–0.7 percent more than the cultivar Volzhanin and 0.3–0.9 percent less than the cultivar Predgorny.

Summing up according to the content of dry matters and starch, we conclude that the studied cultivars according to these indicators can be referred to table cultivars.

5.4 CONCLUSIONS

1. Studying the influence of planting time of various potato cultivars on indicators of their morphological development, efficiency, and quality revealed that for each cultivar the choice has to be individual and applicable to concrete climatic zones.

2. In conditions RNO-Alania for cultivars Vladikavkazsky and Predgorny, the optimum time of planting is the third decade of April (the second planting).
3. Earlier cultivars at early planting escape the heat of July period and because of this the tuber formation stage occurs under optimum conditions.

KEYWORDS

- **Marketability**
- **Matter**
- **Planting**
- **Potatoes**
- **Productivity**
- **Starch**
- **Tube**

REFERENCES

1. Pisarev, B. A.; Potato productivity depending on the type of fertilizers and time of planting. Pisarev, B. A.; Konovalova, L. N.; Bassiev, S. S.; *Potatoes and Vegetables.* **1993**, *5,* 21 p. (in Russian)
2. Bassiev, S. S.; Time of Planting and Fertilization of Various Potato Cultivars. Collection of Scientific Works. "Biological Diversity and Ecological Monitoring in RNO-Alania." Vladikavkaz: North Ossetia State University; **2000**, 41–43 p. (in Russian)
3. Dmitriev, Z. A.; Optimal Time and Density of Planting. Dmitriev, Z. A.; *Potatoes and Vegetables.* **1985**, *2,* 15–17 p. (in Russian)
4. Dospekhov, B. A.; Methods of Field Experiment. Moscow: **1985**, 352 p. (in Russian) (in Russian)
5. Bassiev, S. S.; Yielding Capacity and Quality of Various Potato Cultivars Depending on Agrotechnical Methods on Sandy Derno-Podzolic Soils of the Nonchernozem Zone. Moscow: Autoabstract of Thesis of Agrarian Sciences; **1993**, 26 p. (in Russian)
6. Lorkh, A. G.; Dynamics of Accumulation of Potato Yield. Moscow: "Selkhozizdat", **1948**, 190 p. (in Russian)
7. Kiryukhin, V. P.; Potatoes Physiology. Physiology of Potatoes. Red, N. S.; Bacanov Moscow: "Kolos"; **1970**, *C,* 27–47 p. (in Russian)
8. Prokoshev, S. M.; Potatoes Biochemistry. Moscow: "Kolos"; **1947**, 322 p. (in Russian)

CHAPTER 6

MINERAL NUTRITION AND POTATOES EFFICIENCY IN CONDITIONS OF RNO–ALANIA

SOLTAN S. BASSIEV, DIANA P. KOZAEVA, MARINA V. KATAEVA, and MAGOMED D. GAZDAROV

CONTENTS

6.1   INTRODUCTION

The most powerful factor of impact on growth and development of plants is ensuring the continuous nutrition with all necessary elements, including microcells. Physiological laws of "minimum" and "factors equivalence" testify that the yield is limited by an element that is minimum, but the maximum efficiency is reached when providing the other elements in the necessary proportions, that is, at a balanced nutrition [1].

Use of derno-podzolic soils not resupplying the organic substance and nutrition elements significantly reduces their fertility. Long application of only mineral fertilizers has often negative impact on the chemical, physicochemical, and biological properties of soils [2–5]. Preservation and reproduction of fertility of soils is an important task, especially in modern conditions of maintaining agricultural production when reducing the introduction of organic and high cost of mineral fertilizers.

6.2   THE AIM OF THIS WORK

In this regard, we conducted an experiment to identify the responsiveness of various potato varieties on the level of plant mineral nutrition in mountain zone on RNO–Alania.

Village Kurtat is located in the mountain meadow subalpine zone of Fiagdon hollow lying within the Northern slope of the Central Caucasian Mountains between Rocky and Bokovoy ridges, at a height of 1,450 m above sea level.

Mountain meadow subalpine derno-podzolic soils with the largest arable plots are the main soil difference.

Depending on soil-forming types, the granulometric composition of mountain meadow subalpine soils is not the same. The soils formed on sandstones have easier granulometric composition than soils formed on slates and limestones.

If on granites and sandstones, soils are easily accessible or even sandy loam (the content of physical clay content in the upper level varies from 19.3 to 20.8%), on slates and limestones, the soils are mean and hard loamy (the content of physical clay content in the upper level varies from 22.1 to 41.6%) [3].

Mountain meadow subalpine soils are characterized by the high content of organic substance. In the turf level of subalpine soils, the accumulation of turf is not observed, but due to undecomposed plant remains the content of organic matter is quite high— 31 percent [3, 4].

According to K. Kh. Byasov et al. [4], mountain meadow subalpine soils, despite the high content of gross phosphorus (0.32–0.35%), are very poor in their labile forms. In the turf level, the content of labile phosphorus varies within 2.8–2.4 mg/100 g soil. All soils, irrespective of soil-forming types, are highly provided with potassium— from 30.3 to 51.0 mg/100 g soil. The upper humic levels contain 0.62–1.17 percent of total nitrogen, whereas hydrolyzed nitrogen—from 6.44 to 6.72 mg/100 g of the soils. The content of humus in the upper level is 6.7 percent. Soils of the plot have subacidic reaction of the soil medium (pH—4.9–5.2) [3].

It should be noted that, despite the specifics of mountain soils in Fiagdon hollow (high rubbliness, close shingle beds and soil-forming material, high washing ability, weak water-retaining characteristic), they completely provide plants with necessary amount of nutrients, moisture, and air. This points to the possibility of crop cultivation.

The considered belt is especially suitable for improvement of seed and food potato diseases [3, 4].

6.3   MATERIALS AND METHODOLOGY

The sum of temperatures for the vegetative period varies within 1,880–2,600°C. Length of the frost-free period reaches 160–170 days. Relative air humidity is usually high during the whole year, 85–89 percent; however, in July–August, it can decrease to 50 percent, sometimes lower [4].

This study was conducted in 2012–2013 in the mountain zone of RNO–Alania on the mountain meadow soils covered with pebbles (1,450 m above sea level, branch of the chair of plant growing of the agronomical faculty, Gorsky State Agrarian University, village Kurtat in Alagir region).

Experience was gained in 2012–2013 according to the following scheme: The levels of mineral nutrition in different phases of development—appearance of shoots, butting, flowering, and fading of potato top mass were determined. Structure of tuber yield in various varieties, dry matters, and reducing sugars in various potatoes depending on mineral nutrition were determined. The reaction on different chemical fertilizers on the following varieties of potato was tested: Volzhanin; Udacha; Bars. We added fertilizers in doses: $N_{32}P_{32}K_{32}$; $N_{48}P_{48}K_{48}$; $N_{64}P_{64}K_{64}$. In a control variant fertilizers did not added.

The experiment was conducted in a fourfold frequency in a total plot area 28 m^2, record plot, 25 m^2. Preceding crop was winter wheat.

Seed material represented a tuber first reproduction weight of 60 and 80 g. Planting was carried out by hand according to the scheme 75 × 30 cm on ridges. Mineral fertilizers such as "nitroamophoska" were introduced in spring before reformation of ridges. Care for plantings is standard: interrow loosening was performed while emerging weeds and soil panning. Preparations Karate and Ridomil-Gold were used against the Colorado potato beetle and potato blight.

6.4   RESULTS AND DISCUSSION

The field experiment was conducted by Dospekhov's method (1985). All observations were done by the methods of the All-Russian research institute of plants protection (1990) and the All-Russian research potato institute (1967 and 1989). Yield record was carried out using the continuous weight method.

Researches results: Researches showed that the approach for the phases of potato development depends on the biological features of varieties, agrotechnical methods, and developed weather conditions during the period of vegetation.

B. A. Pisarev and G. A. Galkin's data [6, 7] testify that the duration of variety of vegetation and time of its passing the separate phenological phases greatly depends on yearly weather conditions. Further development of phases and passing the interphase periods greatly depend on the timeliness of sprouts, and it implies the regularity of plants nutrition and yield accumulation.

TABLE 6.1 Influence of mineral nutrition on phases of potato plant growth (days after planting) 2012–2013

Experience variants	Phase of growth and development			
	Sprout emergence	Budding	Flowering	Wilting
Volzhanin				
Control (without fertilizers)	21	39	48	100
$N_{32}P_{32}K_{32}$	21	42	52	102
$N_{48}P_{48}K_{48}$	21	45	53	103
$N_{64}P_{64}K_{64}$	21	48	55	109
Udacha				
Control (without fertilizers)	19	36	45	96
$N_{32}P_{32}K_{32}$	19	38	47	98
$N_{48}P_{48}K_{48}$	19	39	47	98
$N_{64}P_{64}K_{64}$	19	41	51	100
Bars				
Control (without fertilizers)	22	38	47	98
$N_{32}P_{32}K_{32}$	22	41	50	100
$N_{48}P_{48}K_{48}$	22	43	51	102
$N_{64}P_{64}K_{64}$	22	46	53	106

Weather conditions of the second year had a more favorable impact on sprout emergence. Therefore, in 2013, we noted that the early ripening variety Uducha emerged 3 days earlier than the variety Bars and 2 days earlier than Volzhanin. Sprout emergence did not depend on the level of mineral nutrition, but this is not true for subsequent growth phases and interphase periods. Thus, with a rise in the dose of mineral nutrition, interphase periods, and number of days of sprout emergence was increased compared to the next phase. With increasing dosage rate of fertilizer the interphase periods increased as well. For example, in the variant with maximum nutrition level the phase of full dying off came at 109, 100, and 106 days, and this means that varieties shifted from the early-ripening to middle-early and mid-ripening group.

The bush of potato plant consists of several rather autonomous stalks, the quantity of which is a varietal feature. As a rule, between a number of stalks and formed tubers, there is a positive dependence; however, a direct link is often lacking and generally this sign (stalk quantity) is classified as the varietal features [2, 8].

Considering data of our researches, we can note that the variety Bars had the maximum plant height, the variety Udacha conceded to it but not much, but the variety Volzhanin had the minimum height (Table 6.1). Bars had maximum signs according to the top mass and leaf area.

Thus, with increase in the dose of mineral nutrition, plant height, top mass, and area of assimilatory surface also increased. According to Table 6.2, the increase in the dose of fertilizers has no essential impact on stalk quantity. Therefore, we can note that this sign (quantity of stalks per a bush) belongs to varietal features.

TABLE 6.2 Influence of mineral nutrition on formation of potato top mass during flowering phase: years 2012–2013

Variant	Structure of top mass			
	Plant height (cm)	Stalk quantity (per bush)	Top mass (g/bush)	Leaf area (m²/bush)
Volzhanin				
Control	42	3.2	274	1.8
$N_{32}P_{32}K_{32}$	47	3.2	296	2.0
$N_{48}P_{48}K_{48}$	51	3.0	320	2.4
$N_{64}P_{64}K_{64}$	55	3.4	352	2.3
Udacha				
Control	46	4.1	310	2.0
$N_{32}P_{32}K_{32}$	56	4.0	420	2.3
$N_{48}P_{48}K_{48}$	58	3.9	436	2.3
$N_{64}P_{64}K_{64}$	65	4.2	512	2.5
Bars				
Control	48	4.2	320	2.1
$N_{32}P_{32}K_{32}$	60	4.1	412	2.5
$N_{48}P_{48}K_{48}$	65	3.9	486	2.4
$N_{64}P_{64}K_{64}$	69	3.7	524	2.8

In the variety Volzhanin, on the variant, with the application of $N_{64}P_{64}K_{64}$, plant height increased by 13 cm in relation to control, the top mass—78 g/bush, the leaf area—0.5² m/bush. In the varieties Udacha and Bars, on the same variant, the increase of 19 cm, 202 g/bush, 0.5² m/bush, 21 cm, 204 g/bush, and 0.7² m/a bush was noted. With increasing doses rose all indicators structure of aboveground mass of potato plants of all studied sorts (Table 6.3).

TABLE 6.3   Structure of tuber yield in various varieties depending on mineral nutrition, years 2012–2013

Experiment variants	Yield (t/ha)	Yield structure		
		Marketability (%)	Starch content (%)	Nitrates content (mg%)
Volzhanin				
Control	10.2	78	13.5	23
$N_{32}P_{32}K_{32}$	13.6	81	13.8	56
$N_{48}P_{48}K_{48}$	15.9	86	14.1	68
$N_{64}P_{64}K_{64}$	19.9	90	13.6	129
HCP_{05}	1.08			
Udacha				
Control	12.5	82	14.6	45
$N_{32}P_{32}K_{32}$	16.5	85	16.4	65
$N_{48}P_{48}K_{48}$	20.1	88	16.9	77
$N_{64}P_{64}K_{64}$	25.6	92	15.2	128
HCP_{05}	1.72			
Bars				
Control	12.6	83	15.9	26
$N_{32}P_{32}K_{32}$	16.8	86	17.8	46
$N_{48}P_{48}K_{48}$	20.9	89	18.5	66
$N_{64}P_{64}K_{64}$	25.8	93	16.6	126
HCP_{05}	1.73			

Data of our researches revealed (Table 6.4) that the increase in productivity and marketability, and also quality indicators (the content of starch, dry matters, and reducing sugars), during the increase in fertilizer dose occurred due to a certain increase in level of mineral nutrition, and then started to decrease. The nitrates content of all the varieties did not exceed the maximum permissible rates (250 mg of %/kg), but it is possible to note that with the rise in the level of mineral nutrition this sign increased. At dose $N_{64}P_{64}K_{64}$, the content of nitrates for varieties Volzhanin, Udacha, and Bars were 129, 128, and 126 mg of %/kg, respectively.

The variety Udacha exceeded Volzhanin in productivity (Table 6.4). Its marketability is also high; the content of starch is at same level with standard variety and little concedes to Bars in all signs.

The variety Bars showed itself in the positive light in productivity and qualitative indicators and exceeded the total efficiency of the zoned variety Volzhanin at 2.5–6.0 t/ha.

TABLE 6.4 The content of dry matters and reducing sugars in various potato varieties depending on the level of mineral nutrition

Experiment variants	Varieties					
	Volzhanin		Udacha		Bars	
	Content in tubers (%)					
	Dry matters	Reducing sugars	Dry matters	Reducing sugars	Dry matters	Reducing sugars
Control	17.3	0.65	18.3	0.19	19.9	0.18
$N_{32}P_{32}K_{32}$	18.4	0.36	20.9	0.21	21.3	0.19
$N_{48}P_{48}K_{48}$	16.5	0.76	21.7	0.29	22.4	0.21
$N_{64}P_{64}K_{64}$	15.7	0.78	20.9	0.32	20.9	0.20

Indicator of dry matter content in potato tubers is one of the major factors for chips and French fries production. Many authors agree that the high content of reducing sugars promotes pulp blackening in raw and boiled state, which reduces tuber marketability.

The content of sugars in the studied varieties was within the normal range except the variety Volzhanin that gained 0.35–0.77 mg % on experience: variants due to which boiled potatoes of this variety turned black quicker than others. Less reducing sugars were gained with the variety Bars (0.18–0.23 mg %) that points to its suitability for industrial processing.

6.5 CONCLUSIONS

1. Researches are interested from the point of view of steadily high and ecologically safe harvests in the modern terms of agricultural production while at the background reducing organic fertilizers and their costliness.
2. Mineral fertilizer dose $N_{64}P_{64}K_{64}$, applied on prospective potato varieties, Bars and Udacha, in the mountainous area provide excellent tuber quality and high yield (25.8; 25.6 t/ha) commercial potato, at the level of 93 and 92 percent.
3. Maximum quantity of reducing sugars observed at raised doses of fertilizers.
4. There was growth of maintenance of reducing sugars due to the introduction of enhanceable doses of fertilizers.

5. With the increase of dose of mineral feed, all indexes of structure of above-ground mass of plants of potato increased on all studied sorts.
6. Increase of indexes of quality (contents of starch, dry substance, and reducing sugars) at the increase of dose of fertilizers became true due to the certain increase of level of mineral feed,
7. On reaching a certain level, there was a decline of indexes of quality. The table of contents of nitrates in all variants did not exceed possible norms (250 mg%/kg) maximum.
8. Nevertheless, this index rose with the increase of level of mineral feed. With regard to the dose of $N_{64}P_{64}K_{64}$ on varieties Volzhanin, Udacha, and Bars, their maintenance was 129, 128, and 126 mg of %/kg accordingly.

KEYWORDS

- **Derno-Podzolic soil**
- **Efficiency**
- **Fertilizers**
- **Limestone**
- **Mineral nutrition**
- **Variety**

REFERENCES

1. Elkina, G.; Balance of nutriment elements and potato productivity on podsolic soils. *Bull. Inst. Biol.* **2005,** *4,* 14–19 (in Russian).
2. Bassiev, S. S.; Gazdarov, M. Dz.; Gerieva, F. T.; Tsugkieva, V. B.; and Kozaeva, D. P.; Influence of Mineral Nutrition Level on Efficiency and Quality of Potatoes. Proceedings of Gorsky State Agrarian University; **2013,** *1,* 57–63 (in Russian).
3. Byasov, K. Kh.; Mountain Soils of North Ossetia. Ordzhonikidze: Publishing House "Ir". **1978,** 136 p. (in Russian).
4. Byasov, K. Kh.; Dzanagov, S. Kh.; et al. Soils/Natural Resources of the Republic of North Ossetia-Alania. Vladikavkaz: Publishing House "Project Progress"; **2000,** 382 p. (in Russian).
5. Sutyagin, V. P.; Principles for Formation of Agrophytocenosis Stability in Adaptive and Landscape Agriculture. Tver: Publishing House "AGROSFERA TSAA" *(State Agricultural Academy, Tver)* **2007,** 260 p.
6. Pisarev, B. A.; and Galkin, E. M.; Cultivation of Early Potato. Methodical Materials. Moscow: Publishing House "Kolos"; **1967,** 86 p. (in Russian).
7. Mineev, V. G.; Biological Agriculture and Mineral Fertilizers. Mineev, V. G.; et al. Moscow: Publishing House "Kolos"; **1993,** 415 p. (in Russian).
8. Alsmik, P. I.; Shevelukha, V. S.; Ortel, Kh.; et al. Potatoes: Selection, Seed Growing, and Technology of Cultivation. Minsk: Publishing House "Urojai" *(Harvest).* **1988,** 304 p. (in Russian).

PART II
PLANT BREEDING IN THE WESTERN SIBERIA: AGROECOLOGICAL APPROACHES

CHAPTER 7

A STUDY OF ECOLOGICAL AND BIOLOGICAL POTENTIAL AND FORMATION OF GENE POOL OF CULTURAL PLANTS IN INSTITUTE OF BIOLOGY OF TYUMEN STATE UNIVERSITY

NINA AN. BOME, ALEXANDER YA. BOME, ELENA I. RIPBERGER, SERGEY V. ARSENTYEV, and LARISSA I. WEISFELD

CONTENTS

7.1   STRATEGY OF SUSTAINABLE CROP PRODUCTION DUE TO CLIMATIC FACTORS OF THE TYUMEN REGION

The square of territory of Siberia more than 9.6 million km^2, which is 57 percent of the total area of Russia, is home to 16.1 percent of the population. Agro climatic potential is 0.56–0.58 in Western Siberia, Eastern Siberia 0.52–0.54 in the Trans-Baikal region, and the Republic of Khakassia, Tuva, Yakutia, 0.46–0.48 (with averages of 1.0 Russia) [1].

Tyumen region, on the one hand, is a huge-scale oil and gas complex, and on the other is clearly inferior to other regions by bioclimatic potential and development of agricultural land. The lack of heat and moisture during some periods of vegetation has an adverse effect on the level and stability of yields. However, it is well known that the country retains its independence in the event that the ratio of imports to domestic consumption of food and agricultural raw materials does not exceed 20–30 percent. In connection with this, a top priority continues to be agriculture on a sustainable basis, providing valuable opportunity to supply the population with food products of domestic production.

The components of sustainable crop south of the Tyumen region are the selection of plants that are resistant to environmental factors limiting the creation of new highly adaptive varieties, differing in complex breeding and valuable attributes, their rapid reproduction, and proper environmental placement [2].

The human diet due to plant contains an average of 88 and 70 percent energy protein. The number of cultivated plant species evaluated is 5,000, with a relatively widely used plants around 1,200; but more than 90 percent of the energy and protein are produced by cultivated species totaling upto 25–30 [3].

The bases of food security and bio resource qualitative improvement of the environment are of plant genetic resources. Among the many causes leading to the depletion of the plant species are environmental degradation, natural disasters, industrialization, and urbanization [4]. The share of closely related varieties in many growing regions, including the Urals and Western Siberia regions, has increased dramatically and is more than 50 percent [5].

Climate change, manifested by the increase of the average surface air temperature by 0.7 °C in all the continents in the last quarter of the twentieth century [6], indicates the need for adjustments in strategic selection of species and varieties of cultivated plants. It should be noted that Russia at the same time is warmer by 1.5°C [5]. The most intensive the process is in almost the entire European part of Russia, in the south of Western Siberia, Baikal, Trans-Baikal region, and the north-eastern Yakutia [7, 8].

The statistical analysis of the Tyumen hydrometeorology center on environmental monitoring for the period 1993–2012 revealed that of the moisture content of 20 analyzed plant-growing seasons, only four periods (20%) were close to the long-term average value. Moisture deficit was observed for eight (40%) field seasons, with the amount of precipitation ranging from 163 to 276 mm at a rate of 281 mm (minimum rainfall in 1997 and 2012). The share of wet years was 40 percent by varying the amount of rainfall data in the range 322–423 mm.

The daily average air temperature in the nine years of age (45%) exceeded the norm, and only 2 years have been cold. Growing seasons can be divided into three groups: (1) close to the long-term average value for heat and moisture (1993, 1996, and 2001); (2) cold and wet (1999 and 2002); and (3) dry and hot (1998, 2003, 2005, 2010, 2011, and 2012).

These and other factors cannot affect the change in the structure and function of ecosystems. In this regard, the study of plant genetic resources and the formation of collections for specific conditions are of paramount importance. The concept of gene bank storage of plant genetic diversity has been laid and developed in Russia, N.I. Vavilov. At present, according to FAO, the world has 1,750 gene banks, including more than 7 million plant specimens. Of these, there are five main genetic banks in the United States, China, India, Russia, and Japan. The main source of donor genes for the selection of domestic and world are collections of All-Russian Research Institute of Plant Industry, N.I. Vavilov (VIR), with more than 320 thousand of cultivated plants and their wild relatives. This represents 54.1 percent of the gene pool of the ex-Soviet Union countries.

The undeniable role in the study and conservation of plant resources belong to the experimental station, branches, and strong points VIR located in different eco-geographical zones of Russia, including Tyumen outpost at the Institute of Biology of the Tyumen State University in 2005 [9].

Tyumen region on the hydrological and climatic conditions can be divided into the following three major areas: (1) abundant moisture with a lack of heat (the southern boundary of the zone extends between the Khanty-Mansiisk and Tobolsk); (2) excess moisture with lack of heat (the southern boundary extends beyond Tobolsk); and (3) optimal balance of humidity and heat (the southern boundary follows the line Yaluto-rovsk—Ishim—Omsk) [10].

All landed funds for use are divided into two parts: (1) the northern and (2) the southern. The northern part of the district includes two nations—Khanty-Mansi and Yamal-Nenets. Agricultural plants are grown mainly in the south of the Tyumen region.

In general, the soil and climatic conditions of the southern part of the Tyumen region is sufficient for the cultivation of crops, including major cereals—wheat, oats, and barley. At the same time, a large contrast of soil and climatic characteristics in space and in time lead to a significant variation of the complex biotic and abiotic environmental factors. Stress conditions (drought, excess moisture, harsh winters, and reduced high temperature) may occur in a single or several years. All these reveal the importance of selecting varieties that can favorably interact with the environment and ensure high productivity.

Variety is not only a means of increasing the yield, but is also a factor, without which it is impossible to realize the achievements of science and technology [11]. Identification and protection of donors and sources of selection and valuable features and properties, as well as all plant biodiversity, is very important especially with changing technology, extreme situations, and in different environmental conditions.

7.2   STUDY AND CONSERVATION OF CULTIVATED CEREALS IN THE COLLECTION FUND

The ratio of species in the collections formed to a large extent determined by their relevance. In the world, large areas (600–700 million ha) are occupied cereals, and the leading role belongs to wheat (spring and winter).

In the Tyumen region, for the last few years, of the total sown area of crops in spring, wheat average is about 63 percent. We studied the collections of much of the cultural cereals presented: 346 samples of spring wheat, 319 samples of winter wheat, 325 samples of barley, and 53 of oats (a total of 1,043 samples).

Geographical analysis of these collections has revealed that they reflect the diversity of natural factors and the level of breeding achievements of 57 countries of the world (Europe, Asia, North America, South America, Africa, and Australia), and 33 subjects of the Russian Federation, located in all regions.

The uniqueness and the representative character of the genetic structure of collections of crops reflect the chosen species, subspecies, and more than 120 botanical species of plants. The study of the collections is carried out for sample identification, description of their morphological and biological properties, grain quality, and environmental test. Part of the seeds are annually sent to the departments of VIR.

Collections are investigated in the bio station "Lake Kuchak" by Tyumen State University, and their laboratories of the Department of Botany, Biotechnology, and Landscape Architecture with infectious and provocative backgrounds. The signs are chosen taking into account the constraints of the agricultural area of the Tyumen region: early maturity, resistance to drought, disease, logging, salinity, low temperatures, resistance to winter, the content, and quality of gluten.

A changing climate may increase the number of plant diseases caused by warming and therefore of special value samples with resistance genes. It is known that tolerance of resistant varieties is unstable over time due to the emergence of new virulent races of pathogens. Therefore, screening collection fund for resistance to phytopathogenic fungi—agents of rust, powdery mildew, and blights of various etiologies in our research is carried out continuously.

There is a need to develop technologies for the safe and long-term conservation of plant genetic resources due to various terms of longevity of seeds: microbiotics (less than 3 years), mezobiotic (3–5 years), and macrobiotic (over 15 years). In cases of frequent reseeding, plant specimens may give rise to biological contamination or technical errors that lead to changes in the correlation among biotypes in polymorphic systems and loss of valuable genotypes. At low positive temperatures (+4°C), adopted in the world, the seeds remain viable for a long time. Recently, shallow freezing and cryopreservation (−196°C) have been studied as more long-term methods of storing seeds.

The Institute of Biology has a modern refrigeration equipment, which allows for a rapid storage of plant material and microorganisms at temperatures of +4 and −10°C and longer—at temperatures of −80 and −150°C. Before putting the seed for long-term, low-temperature storage, it is necessary to constantly check their moisture content, germination, and phytopathological state. Storage is done in a sealed package. During the storage of seed is viability monitoring. The diversity of the collection of the material provided for its perpetuation and management can provide a significant contribution to the fight against environmental pollution and biodiversity conservation areas.

7.3   EXTENSION METHODS OF GENETIC DIVERSITY OF CULTIVATED PLANTS

7.3.1   HYBRIDIZATION

One of the primary tasks of spring wheat is the creation of original material with a high ecological plasticity, using modern methods, one of which is hybridization.

For hybrids obtained by crossing five varieties of spring wheat incomplete diallel scheme, graduate E. I. Ripeberger, in the first generation (F1) found a less pronounced susceptibility to powdery mildew (*Erysiphe graminis* DC.). Dividing the sample into groups of sustainability revealed 10 percent of hybrid signs of injury. Among the starting grades, a large proportion (40%) had a very low resistance to phytopathogenic fungi of hybrids dominated combination with an average degree of lesion.

With regard to the spotty (*Alternaria* sp., *Helmintosporium* sp.), 50 percent of hybrid showed a high level of stability. While the initial varieties group was only 20 percent. Among the dominating varieties, the group with a very low and medium resistance, including hybrids, accounted for 10 and 20 percent, respectively.

According to the results of the studies conducted in 2010, 2011, and 2012, in terms of resistance to powdery mildew stood a hybrid form of crossing ♀ Hybrid × ♂ Lutescens 70. Very high resistance to spotted was possessed by hybrid forms ♀ SKENT 3 and ♂ SKENT 1, ♀ Hybrid × ♂ Lutescens 70, ♀ Hybrid × ♂ SKENT 3, and ♀ Cara × ♂ SKENT 3, the initial variety Lutescens 70 [12-13].

Dedicated by us the immune samples wheat were several unique forms that are both resistant to several most harmful pathogens - pathogens of fungal diseases. They can be used in breeding as potential donors of resistance to develop new varieties for the competitive global market [13].

The basic foundation for the identification of the causative agent, the study of developmental biology, and the creation of infectious backgrounds is a collection of pure cultures of pathogenic fungi and mycological herbarium. The collection includes more than 150 strains, including the most damaging of representatives of the genera *Helmintosporium, Fusarium, Alternaria, Sclerotinia*, and others is important from both a theoretical and practical point of view.

Yu.B. Trofimova obtained new data on the biology of pathogens of various plant diseases. Thus, it was found that the effect of *Fusarium nivale* CES. depends on the temperature. When cultured on potato glucose agar, beginning of the growth of the fungus (on day 8), sporulation (on the 56th day), and formation of colonies (90 mm—42th day) were observed at a temperature of 5°C. The defeat of snow mold collection samples of winter wheat in some cases led to a thinning in the other and to the complete destruction of plants. Highlighted during the course of many years of research, sources of resistance to this and other phytopathogenic fungi are a valuable raw material for adaptive selection [14, 15].

In studies performed at the Department of Botany, Biotechnology and Landscape Architecture with rye, O.A. Buldyaeva convincingly showed varietal differences in terms of susceptibility to the effects of fungi of the genus *Puccinia* Pers. in northern forest-steppe of the Tyumen region [16]. Staring with the appearance of brown rust on plants at earing, maximum development of the disease is at the phase of milk ripeness.

There was a significant linear decrease in the size and weight of the leaves of plants infected with brown leaf rust. On average, the maximum reduction of 12 samples were indicated by bone-dry (up 63%) and crude (43%), the mass of leaves. In general, a complex of traits (length, width, and leaf area), and the largest inhibition of leaf surface, is characteristic of moist, warm, and favorable for disease development growing seasons [17]. There was some fixed dependence between the yield and the degree of development of the disease. Bond strength varied from a high in 2002 ($r = 0.73$) to average in 2003 ($r = 0.65$) and lowest in 2004, 2005, and 2006 ($r = 0.10$; $r = 0.20$; and $r = 0.26$, respectively), which we attribute to the peculiarities of hydrothermal regime of the study periods (average daily air temperature, effective temperature, and rainfall).

7.3.2 EXPERIMENTAL MUTAGENESIS

Mutation process in combination with hybridization provides heterogeneity of plant material and maintains it for a long time. I.I. Shmalgauzen [18] put forward the idea of "mobilization reserve of genetic variation," which appears in the extreme environmental conditions. It is no accident that in the center of contemporary projects are genetic variation and genetic diversity.

Our researches in experimental mutagenesis study conducted in 1972—initially in the laboratory breeding of the Research Institute of Agriculture of the Northern Trans-Urals, and then in the laboratory of biotechnology and microbiology of the Tyumen State University. Comparison of long-term data on experimental mutagenesis on self- and cross-pollinating plants have established general rules in their sensitivity in the first generation after seed treatment (M_1) and mutability, that is, the frequency of mutations in the second generation (M_2) under the influence of physical and chemical factors [19].

We studied the effect of alkylating compounds phosphemid (phosphemidum, synonym phosphazin)—di(ethylene imid)-2-pyrimidylamidophosphoric acid [20]. The phosphemid is interesting that, on the one hand, it has two groups of well-known mutagen of ethylene imine and, on the other hand, it has a pyrimidine base [21, 22]. This difficult structure is a mutagen, and in can join in DNA during the synthesis of chromosome due to the pyrimidine base. Phosphemid was synthesized in 2013 by Prof. Eugeny V. Babaev in Chemical Laboratory in Chemistry faculty of Moscow State University.

Research was conducted on varieties and hybrids of spring and winter wheat forms. Cultivars of winter wheat taken from the collection fund: Kroshka (k-63050, Krasnodar), Bezenchukskaya 1 (k-4278, Samara Region), and Gunistan (k-64283, Krasnodar). According to the results of years of research (2007–2012), varieties stood out for a number of valuable traits in breeding and breeders can be offered as a starting material [23, 24]. There is a need for further work with the material based on the seed, caused by mutagenic treatment available along with advantages and disadvantages (poor plasticity and environment, and as a result, unstable yield data).

In spring wheat, the subjects of study were two varieties of different eco-geographical origin: Cara variety of worldwide collection (k-64381) Mexican breeding and variety SKENT 3 of Russia, as well as a hybrid of the fourth generation (F_4), derived from crosses of these varieties. The hybrid combination has passed a comprehensive evaluation to the selection of forms, having useful features.

The test was conducted in 2013 to obtain mutants of the first (M_1) after exposure to a mutagen generation. The dry seeds of spring and winter wheat were treated by phosphemidum at concentrations of 10^{-2} and 10^{-3} M by the method described by L. I. Weisfeld [22]. Seeds were soaked in mutagen solutions at room temperature (20–22°C) for 4 hr. After treatment, seeds were then washed under running tap water for 60 min. In the control was taken distilled water. Seeds of spring wheat after treating with a mutagen were studied under field conditions during the growing season in a 4-fold repetition in May-August 2013. In the laboratory experiment, seeds of winter wheat were placed in Petri dishes on filter paper of 50 pieces in each, in 4 replicates. Seeds were germinated in a thermostat at 25°C. The vigor and laboratory germination of seeds and seedlings of morphometric parameters were determined. In August 2013, a field experiment on samples of winter wheat was conducted. The number of seeds sown on each variant was 200 units.

The main objective of this work on chemical mutagenesis is to increase the genetic diversity and create a collection of mutant forms of wheat by mutagenic treatment of seeds. Mutational variability was studied on seedlings in the laboratory and on plants, in the field.

In our previous experiments [25–27], in several species of cultivated plants using gamma rays, NMM, NEM, EI, DMS, DES, the observed polymorphism pronounced resistance to its complete absence. Under the same exposure modes on mutagens (gamma rays, and the concentration of chemical mutagens), some samples very rapidly decreased indicators of seed viability, growth, and development of plants in the first generation after mutagen treatment M_1. These include soft wheat—Strela, sainfoin sandy SibNIIK 41, clover Falenski 1, amaranth—sample k-147. Other samples of morpho-physiological parameters did not change significantly and remained at the level of control. This group consisted of soft wheat Haruhikari, sainfoin sandy Sib-NIIK 3274. Some objects manifest the effect of stimulation of growth processes: from soft wheat Milturum 553 and sainfoin sandy Phlogiston to amaranth k-40197, k-47. The hybrid forms of spring wheat had characteristic large stability in relation to the action of mutagens as compared to original varieties.

Under the action of phosphemid, a hybrid form F_4 soft spring wheat Cara × SKENT 3 increased survival compared with control plants after treatment of the first generation during the growing period in 2013. Compared with the best biological stability of the initial variety in the embodiment, hybrid plants at a concentration of 10^{-2}M were higher by 40.0 percent. Thus, the results obtained by us at the present time confirm the results of earlier studies.

Hybridization may have created a new type of organism. Progeny studied hybrids have the capacity for repair of mutagenic damage.

Phosphemid had a stimulating effect in the germination of seeds of winter wheat cultivated in podzolic soil, which was found in two varieties of winter wheat collection:

Kroshka (k-63050, Krasnodar) and Bezenchukskaya 1 (k-642787, Samara Region). Germination of seeds in versions with a mutagen exceeded the control by 11.8–22.6 percent. At the same time, the variety Gunistan obtained from the Krasnodar area (k-64283) showed inhibition of the growth processes in the early stages of ontogeny.

Stimulus effect of chemical mutagens by the nature of the display is similar to the effect of low doses of ionizing radiation, which is explained in certain hypotheses.

Toxicological hypothesis of radio stimulation proposed by N.V. Timofeev-Ressovsky [28], according to which the effects of ionizing radiation on cell breakdown products occur, led to an increase in the rate of cell division in the body. At high concentrations, toxic breakdown products are destructive to cells and exciting life activity at low concentrations.

I. A. Rapoport proposed the concept of "dual genetic stimulation" [29–31], which implies the existence of several private mechanisms of genetic stimulation. Stimulation of heterozygote is to increase and as a consequence leads to heterosis body. According to the author, the stimulation can acquire a significant economic efficiency, as it allows reducing the length of time needed to create lines that allow transfer method heterosis for self-pollinated plants.

Looking at this question from the point of view of practice, it should be remembered that the ratio of heterosis and nonhegeterosis contribution of stimulation are different for different species and varieties. Nonheterosis mechanism represents the modification, that is, nonhereditary stimulating effect. It is assumed that chemical mutagens are both genes and the enzyme apparatus defining metabolism, growth, and proliferation of cells.

In our experiments [32], the influence of stimulating doses of chemical mutagens increased the indexes of seed germination, the survival of the plants during the growing season, and yield. The factors determining the variability of symptoms when exposed to a number of mutagens, the frequency and variety of mutations are the specific actions of mutagens and genotypic characteristics of the facility. Differences in the mutability of individual genotypes should be evaluated by the total frequency of mutations induced by the same mutagen in various doses. The reproducibility of the genetic effect by using the mutagenic factors in a specific dose is low because of the stochastic nature of mutational changes.

7.4   OTHER COLLECTIONS OF PLANTS

Along with cultivated cereals, we form a collection of other plant species. In a study of 140 samples of peas from 15 regions of the Russian Federation (70% of samples) and 21 samples from other countries (30%).

Increase of the productivity of symbiotic nitrogen fixation and legume plants and production of high-quality and environmentally friendly raw materials is determined by the effectiveness of biological products on the basis of symbiotic and associative rhizosphere microorganisms [33]. Work is carried out under a joint program with the All-Russian Scientific Research Institute of Microbiology of the Russian Academy of Agricultural Sciences (St. Petersburg).

Department of Botany, Biotechnology, and Landscape Architecture of Tyumen State University has a collection of medicinal plants, in which there are 40 species, obtained both from foreign countries (France, Germany, Latvia, and Switzerland) and the different regions of Russia (Novosibirsk region, Altai region, etc.). In the vicinity of the biological research station "Lake Kuchak" graduate student A.N. Mikhailova collected several wild forms. These forms are unique and are clearly useful for the exchange fund.

To preserve biodiversity and the possibility of production of raw medicinal plants, studies have been conducted on the cultivation of wild forms under simulated conditions. The results of the comparative assessment of the growth and development of plants in culture and in natural habitats suggest this line of promise [34].

The fund has a large department of botanical and geographical diversity of ornamental plants, totaling 159 specimens. Of particular value for cultivation in the urbanized areas of northern cities are samples that form viable seeds. Dedicated samples are suitable for use in landscape design, which allows the company to provide specialized Tyumen seeds of its own production.

7.5 CONCLUSIONS

The widespread use of samples from the world collection of All-Russian Scientific Research Institute of Plant Industry of N.I. Vavilov (VIR) and hybrid and mutant forms in genetic and breeding research is due to the need to introduce a new culture of dominant genes. This approach will allow the major challenge of the selection process associated with a genetic adaptation of varieties to the conditions of the region, which significantly eliminates the dependence of yield changes in climatic and other conditions.

We conducted the selection and creation of genetic sources of resistance to biotic and abiotic stresses through the use of recombination and mutation variability of plants.

At the present stage, the use of experimental mutations in the selection is important. This is due to the fact that breeding success, which is based on the selection, is closely linked to the presence of the gene pool of spontaneously occurring mutations. However, the fund is depleted of natural beneficial mutations, which leads to a decrease in the probability of a favorable combination of genes and the creation of new genotypes in hybrid combinations. Obtaining mutations artificially significantly faster compared to natural conditions create a variety of genotypes that are the source of recombinants with the new gene interaction.

Practical importance is the completion of the existing Institute of Biology of the Tyumen State University genetic stock to sustainable use of plant resources.

We developed a control system gene pool of plants, including the identification of genotypes, the creation of passport and evaluation databases on the results of eco-geographical and laboratory studies of samples of plants from the world collection of VIR; other research institutions and other samples were obtained as a result of their own funds.

KEYWORDS

- **Collection**
- **Environmental factors that attribute**
- **Gene pool**
- **Genetic resources**
- **Pattern**
- **Sustainability**
- **Variety**

REFERENCES

1. Surin, N. A.; Breeding and Production of Crops in Siberia. Surin, N. A.; St. Petersburg: I International Congress of Russian Grain and Bread; **2005,** 102 p. (in Russian).
2. Bome, N. A.; Selection of Crop Varieties and Methods of Creation for the Extreme Conditions of the North Trans-Ural. Abstract. Dis. St. Petersburg: Doctor of Agricultural Sciences; **1996, 46** p.
3. Zhuchenko, A. A.; Ecological Genetics of Crop Plants and the Problems of Agricultural Sphere (Theory and Practice). Moscow: **2004,** *1,* 688 p. (in Russian).
4. Aleksanyan, S. M.; The Strategy of the World Gene Banks Interaction Under Globalization. Works of applied botany, genetics and plant breeding. Spb.: VIR; **2007,** T. *164,* 11–33 (in Russian).
5. Martynov, S. P.; and Dobrotvorskaya, T. V.; Manifestation of Genetic Erosion in the Dynamics of the Diversity of Russian Wheat. St. Petersburg: Presentation Workshop Vavilov; January 28, **2009** (in Russian).
6. Tyutyuma, N. V.; Theoretical and Applied Aspects of the Study of the Gene Pool of Breeding Value of Cereal Crops in the Arid Conditions of the Lower Volga. Abstract Dis. Astrakhan: Doctor of Agricultural Sciences; **2009,** 44 p. (in Russian).
7. Anisimov O.F., Grigoriev M.N., Instanes A. et al. Assessment report on climate change and its consequences on the territory of the Russian Federation. Anisimov O.F., Grigoriev M.N., Instanes A. et al. Vol. I. Climate Change. Moscow: RosHydromet; **2008,** 228 p. (in Russian).
8. Zamolodchikov, D. G.; Evaluation of Changes Climatogenic Diversity of Tree Species on Forest Inventory Data. Advances in Modern Biology; **2011,** T. *131(4),* 382–392. (in Russian).
9. Bome, N. A.; Integration of research and education (for example, the reference point of the Tyumen research institute of plant industry). Vavilov, N. A.; and Bome, A.Y.; Bome Successes of Modern Science; **2007,** *12,* 88–89. (in Russian).
10. Cherkashenina, E. F.; Agro-Climatic Resources of the Tyumen Region (Southern Part). Leningrad: Gidrometeoizdat; **1972,** 151 p. (in Russian).
11. Efremova, V. V.; Changing the varietal composition agrotcenoze winter field. The Anniversary Edition for the 75th Anniversary of the Krasnodar State Agricultural University. Ephraim, V.; Aistova, Y. T.; and Terpugova, N. I.; Krasnodar: Agro Ecological Monitoring in Agriculture Krasnodar Region; **1997,** 468 p. (in Russian).
12. Ripberger, E. I.; Intraspecific diversity of hybrids *Triticum aestivum* L. for resistance to phytopathogenic fungi. Ripberger, E. I.; Bome, N. A.; and Bome, A. Y.; Fruit and Cultivation of Berries of Russia; **2012,** Vol. *XXXIV,* Part 2, 175–182. (in Russian).
13. Arsent'ev, S. V.; Sustainability of collection samples *Triticum aestivum* L. phytopathogenic fungi in a changing environment. Arsent'ev, S. V.; Bome, N. A.; Kolokolova, N. N.; and Bome, A. Y.; Innovative Development of Agriculture of the Northern Trans-Urals. Proceedings of the regional scientific-practical conference of young scientists. Tyumen State Agricultural University of Northern Trans-Urals; **2013,** 7–14 p. (in Russian).

14. Trofimova, Y. B.; Stability of Winter Rye to Phytopathogenic Fungi Under Different Environmental Conditions. Abstract. Con. Tyumen: The Candidate of Biological Sciences; **2005,** 21 p. (in Russian).

15. Trofimova, Y. B.; Parameters and Severity of Snow Mold Resistance of Winter Rye to Illness. Trofimova, Y. B.; and Bome, N. A.; Journal of Plant Protection. Number 1. Pushkin: St. Petersburg; **2006,** 33–36 p. (in Russian).

16. Buldyaeva, O. A.; The Parameters of Leaves of Winter Rye which are amazed by Rust (in the Northern Forest-Steppe of the Tyumen Region). Buldyaeva, O. A.; Modern High Technologies; **2004,** *1,* 48 p. (in Russian).

17. Buldyaeva, O. A.; Plant Resistance of Winter Rye to the Defeat of Plant Pathogens (for Example, Fungi of the Genus *Puccinia* Pers.) Author Con. Tyumen: The Candidate of Biological Sciences; **2007,** 22 p. (in Russian).

18. Shmalgauzen, I. I.; Factors of Evolution (the Theory of Stabilizing Selection): Print. Second revised and updated. Moscow: Science; **1968,** 451 p. (in Russian).

19. Bome, N. A.; Mutational variability of some species of plants and reparative effect of para-amino benzoic acid. Bome, N. A.; and Bome, A. Y.; Indukovany Mutagenesis Selektsii Roslyn. Bila Churches; **2012,** 53–60 p. (in Russian).

20. Chernov, V. A.; Cytotoxic Agents in Cancer Chemotherapy. Chernov, V. A.; Moscow: Medicina; **1964,** 320 p. (in Russian).

21. Weisfeld, L. I.; Cytogenetic Effect Phosphazin on Human and Mouse Cells in Tissue Cultures. Russian Journal of Genetics; **1965,** *4,* 85–92 p. (in Russian).

22. Weisfeld, L. I.; Chromosomes and mitotic activity by influence of alkylating agent phosphemidium. *Polym. Res. J.* **2013,** *7(1),* 9–22.

23. Tyumentseva, E. A.; The Study of the Reaction of Collection Samples of Winter Wheat on the Weather Conditions on the 1,000 Grain Weight. Tyumentseva, E. A.; and Bome, N. A.; Fruit Growing and Cultivation of Berries in Russia; **2012,** T. XXXIV, Part 2, 334–342 p. (in Russian).

24. Bome, N. A.; Formation of the Leaf Surface Forms of Winter *Triticum aestivum* L. In Different Climatic Conditions. Bome, N. A.; Tyumentseva, E. A.; and Bome, A. Y.; Bulletin of the Tyumen State University Publishing House of the TSU; **2011,** *12,* 132–137 p. (in Russian).

25. Bome, N. A.; The study of mutational variability of hybrids of spring wheat induced by gamma radiation and its selective use. Author Con .... The Candidate of Biological Sciences. Abstract of a thesis of PhD (Biol). Leningrad: VIR; **1980,** 22 p. (in Russian).

26. Bome, N. A.; The action of chemical mutagens and gamma radiation on the perennial legumes. Bome, N. A.; and Komarov, A.; Chemical Mutagenesis in the Selection Process. Moscow: Nauka; **1987,** 174–176 p. (in Russian).

27. Bome, N. A.; Mutational Variability of Three Samples of Barley Induced by Chemical Mutagens. Breeding, Genetic and Environmental problems of Eukaryotes. The collection of proceedings of the Tyumen State University. Ed. Tsoy, R. M. Tyumen: Publishing House of the TSU; 1995, 3–7 p. (in Russian).

28. Timofeev-Ressovsky, N. V.; Biochemical Interpretation of Phenomena Radio Stimulation of Plants. Biophysics; **1956,** *1(7),* 616–627 p. (in Russian).

29. Rapoport, I. A.; Features and Mechanism of Action Super Mutagen. Supermutagens. Moscow: Science; **1966,** 9–22 p. (in Russian).

30. Rapoport, I. A.; Dual Genetic Stimulation Induced Super Mutagens Mutation Breeding. Moscow: Science; **1968,** 230–241 p.

31. Rapoport, I. A.; Shigaeva, M. H.; and Ahmatulina, N. B.; Chemical Mutagenesis. Alma-Ata: Science; **1980,** 317 p.

32. Bome, N. A.; Sustainability of Crop Plants to Environmental Stress. Monography. Bome, N. A.; Bome, A. Y.; Belozerova, A. A.; Tyumen: Tyumen Publishing House of the Tyumen State University. **2007,** 191 p.

33. Savinyh, A. A.; Efficiency of Inoculation of Pea Rhizobia in the Northern Forest Region of Tyumen. Savinih, A. A.; Kolokolova, N. N.; Bome, N. A.; Modern High Technologies; **2007,** *2,* 69–70 p.

34. Mikhailova, A. N.; The study of reactions of medicinal plants to changing environmental conditions on the duration of phenological phases. Mikhailova, A. N.; and Bome, N. A.; Natural and Technical Sciences. Publishing House of the Sputnik; **2012,** *6(62),* 139–144 p.

CHAPTER 8

BASIC PROPERTIES OF SOD-PODZOLIC SOIL OF NORTH TERRITORIES OF TYUMEN REGION

IRINA AL. DUDAREVA and NINA AN. BOME

CONTENTS

8.1   INTRODUCTION

Nowadays, comprehensive and detailed study of the main characteristics, which are designated properties of certain sod-podzol soil for the growth of cultivated plants in adverse agroclimatic conditions of northern territories, are of relevance [1–6].

Tobolsk district is situated in the northern part of Tyumen region, in subboreal forest zone, and occupies 17.222 km². Summer time climate is formed under the influence of cyclones moving from the west. However, intrusion of arctic air causes cooling and frosts in the beginning and in the end of summer period. Anticyclones of the Central Asia enhance climate continentally during winter, which leads to relative severity of the period [7]. Territories of the region are characterized by severe cold winter and short frost-free period. Not only annual but daily sharp temperature fluctuations are observed, especially during spring. Climate instability is related to unhampered intrusion of arctic air masses from the north and dry air masses from Kazakhstan.

By its hydrological-climatic conditions, Tobolsk district belongs to highly humid zone and zone of insufficient heat supply [8].

Climatic conditions have essential influence on soil formation process and determine geographical and physicochemical uniqueness of soil cover. This region differentiates by the great cultivar of soils. Main soil types are as follows: floodplain, podzols, sod-podzols, gray forest soils, alluvial-meadow soils, boggy soils, black soils, sodium soils, ash gray soils, and loessial loam.

Investigations revealed great spatial and time contrast of edaphic-climatic characteristics in complicated conditions of growth of cultivated plants [2]. That is why it is essential to know the morphological, physical, and chemical peculiarities of soil cover and its ultimate composition for realization of ecological and biological potential of crop cultivar [6, 9, 10].

The aim of this study is to determine the morphological features, physical and chemical characteristics, and elemental ultimate composition of soil from an experimental plot with the account of meteorological factors.

The research was conducted in the period 2009–2011 on the experimental plot in Malaya Zorkaltseva village, Tobolsk district of Tyumen region, according to agroclimatological zoning situated in subboreal region.

Data on air temperature in Tobolsk district for the period 1900–2008 were obtained on "Joint hydrometeorological station of Tobolsk district." Analysis of multiannual data revealed that average monthly air temperatures in the period from November to March has negative values, and from April to October—positive; mean yearly air temperature was 0.6°C.

Yearly variation of temperatures is characterized by minimum in January–February (−17.2°C) and maximum in July (+18.6°C). Mean temperature of cold period from November to March is −13.2°C. The coldest months were January and February with minimal temperatures −48.5°C (1964) and −47.7°C (1967). Duration of period with mean daily temperature above 0°C is 190 days and above +5 and +10°C, 157 and 116 days, respectively. According to multiannual data, minimal July temperature is +39.6°C (1901). Spring begins after April 10 with the passage of daily temperature through 0°C, but in the end of May and in the beginning of June snowfalls are possible. However, at times warm and even dry weather (+15 and −23°C) set in because of moving of dry warm air masses from Kazakhstan. Time of passage of mean air temperature through +10°C is accepted to denote the beginning of summer (May 21–June 10).

Characterizing the thermal regime of soils form Tobolsk district, it is possible to note that they undergo rather protracted and deep freezing in winter; slow thawing and warming of soils is common in spring.

By the results of profiling, it was established that soil of the plot is well-cultivated residual carbonate, sod-podzol on ancient alluvium deposits. The surface is billowy, without erosional features, and profile character is simple unbroken. Signs of textural and structural profile dissimilarity are layers of heavy grain texture up to B-horizon and light grain texture in C-horizon. Parent rock material is sand alluvial deposits of the first terrace above the Irtysh river floodplain. Soil profile composition is the next: Ap (0–38 cm), E (38–48 cm), EB (48–76 cm), Bh,f,al (76–93 cm), and C (93–110 cm).

Productivity of plants cultivated on the soil with high cryogenic load is determined by the level of warm and water supply during their growth and development. The object of research in our field experiment was soft summer wheat.

During the period 2009–2011, vegetation seasons considerably differed both between them and with mean multiannual values. Observed fluctuations of mean daily air temperatures were from +10.8°C (May) to +17.2°C (July) in 2009, from +8.9°C (September) to +17.5°C (July) in 2010, and from +10.8°C (May) to +18.0°C (June) in 2011. Minimal daily temperatures were observed in May 2009 and 2011 and in September 2010; maximal temperatures in July 2009 and 2010 and in June 2011. It is possible to characterize the years of the research as warm with sharp fluctuations of daily and monthly temperatures. The overage above multiannual data made in 2009—2.8°C, in 2010—0.5°C, and in 2011—3.2°C. Sum of active temperatures was above the normal (1,500–1,700°C) at the average for the whole period on 219°C and were 1977.4°C in 2009, 1855.4°C in 2010, and 1925.3°C in 2011.

Precipitation total during plant vegetation was close to norm in 2009 and in 2011 and reached 311.3 and 358.2 mm each year, respectively. The year 2010 vegetation period belongs to critical by precipitation (221.9 mm), which is lower than norm at 73.1 mm. Months characterized by shortage of moisture were determined: July and September 2010 (precipitation totals were 19.9 and 30.0 mm, respectively); May, August, and September 2011 (precipitation totals were 8.5, 46.6, and 39.5 mm, respectively). In some periods, minimal precipitation values were noted: 2009 (July—91.4 mm, August—89.6 mm), 2010 (July—61.4 mm, August—78.7 mm), 2011 (June—162.5 mm, July—101.1 mm).

Laboratory analyses were made on the basis of accredited laboratory "Ecotoxycology" (POCC RU. 0001.516420) of Tobolsk complex scientific station, Ural division of the Russian Academy of Sciences.

Sample collection for the research was made by soil sampling tube according to Russian state standard (GOST) 2816–89, Ruling Documents (RD) 52.18.156-99, GOST 17.4.3.01.

Soil moisture is an important characteristic for passing of ontogeny stages by organism from the moment of seed germination. Moreover, this characteristic has immediate impact on soil chemical composition because of having influence on transition of chemical elements from immobile forms to mobile forms.

Soil moisture regime for certain period was determined according to GOST 28268-89 during vegetation (May, June, July, August, and September). Moisture content in the soil (39.4%) was sufficient and favorable for summer wheat seed germination and sprout formation in June 2011. Soil moisture was low in 2009 and 2010 (11.7 and 13.6% respectively), which was reflected by field germination rate and biological resistance of plants. The important period in water consumption is thought to be booting and ear formation stage, that is, period of reproductive organ formation, which comes to be in July in our research. Maximally hard conditions for wheat were observed in July 2010, which was characterized by precipitation deficiency on the background of increased air temperatures. Soil moisture during this period was no higher than 6.8 percent. Low soil moisture (9.0%) was observed in August 2011, when milk stage of grains took course and plants consumed 20–30 percent of all moisture during vegetation period. Warmth and water regime influenced the soil chemical properties, the degree of mobility of different elements, and the plant's ability to consume them through its root system.

Soil acidity is stated by the negative logarithm of the hydrogen ion concentration pH. This characteristic determines the availability of chemical elements for plant organism. It is worth mentioning that amelioration of sod-podzolic soils leads to changes in the qualitative composition of organic matter, decrease of fulvic acids composition, and increase of lime humates. At the same time, as a result of amelioration, the base exchange capacity increases and composition of exchangeable cations changes: increasing of consumed Ca^{++} and Mg^{++} and decreasing of exchange H^+ and Al^{+++}.

Reaction of soil solution markedly changes because of saturation of soil by Ca^{++} and Mg^{++} cations: pronounced acidity, which is characteristic to virgin soils, gradually replaced by subacidic and sometimes neutral and weakly alkaline reaction. Biological activity of soil microbial flora—nitrate bacteria and azotobacter—which does not occur in virgin soils and weakly cultivated soils or occurs in very fractional amount, intensifies because of it [11].

8.2   MATERIAL AND METHODOLOGY

Soil of the experimental plot belongs to weakly alkaline type and has medium pH 7.70. It is known that nutrients and chemical elements for wheat plants will be available under a pH range 6.0–7.5. If pH level is lower, key nutrients will be either less available or become toxic for plants. Therefore, soil pH of experimental plot can be referred as satisfying to requirements of the culture.

Dry residue (solid residue) is a characteristic of soil salinity; it is determined by the ratio of anions and cations in the soil solution. In normal conditions, it cannot exceed 0.30 percent; in soil samples, it is equal to 0.35 percent. Salinity is determined by salt content in soil solution. Salts need to be formed mostly by sodium, calcium, and magnesium cations with chloric and sulfuric anions. Potash cations, bicarbonate, carbonate, and nitrate anions can comprise the insignificant part. Thus, results by dry residue in soil, which were obtained in this experiment, allow drawing a conclusion that anion and cation amounts are optimal and they are the main compounds of the soil in current agroclimatic conditions.

8.3   RESULTS AND DISCUSSION

The amount of anions in soil is considerably less than the amount of cations. Anion–cation balance is shifted toward cations (FIGURE 8. 1-3).

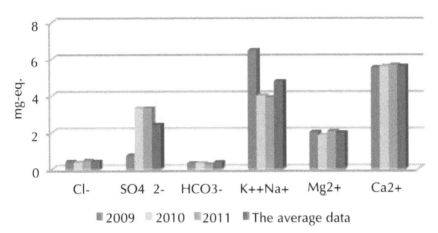

FIGURE 8.1   The content of anions and cations in the soil of experimental site (Cl^-; SO_4^{2-}; HCO_3^-; $K^+ + Na^+$; Mg^{2+}; Ca^{2+}).

On the basis of the obtained data on biogenic substances, it may be concluded that the nitrogen in the soil is presented in three forms: ammoniacal (NH_4^+), nitrate (NO_2^-), and nitrite (NO_3^-); there is a sufficient amount of nitrogen in the soil (Figure 8.2).

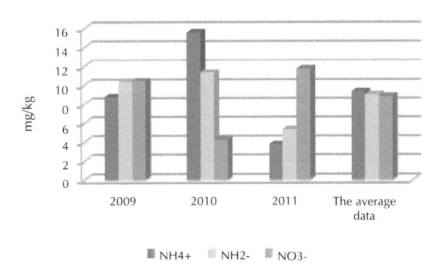

FIGURE 8.2   Nutrients in the soil of experimental site (NH_4^+; NO_2^-; NO_3^-).

Phosphorus is available in mobile forms $H_2PO_4^-$ и HPO_4^-; it is contained in the soil in large amount. Its accumulation is conspicuous over time: the maximum was observed in 2011 and the minimum in 2009 (Figure 8.3).

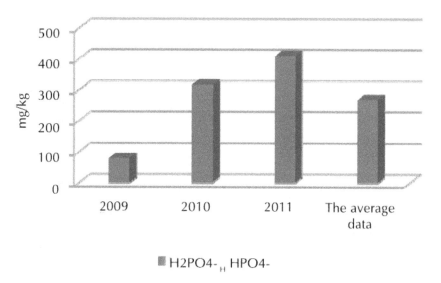

H2PO4- и **HPO4-**

FIGURE 8.3   Mobile forms of phosphorus in the soil of experimental site ($H_2PO_4^-$ и HPO_4^-).

The humus content (I.V. Tyurin method in modification of V.N. Simakov) in the soil is not high and it varied slightly: year 2009—1.86 percent, year 2010—1.45 percent, year 2011—1.76 percent.

The total content of the chemical elements (As, Ca, Cd, Co, Cr, Cu, Fe, Mg, Mn, Mo, Ni, Pb, Sb, Sr, Zn) was estimated in soil samples, as well as their stationary and mobile forms, by atomic emission methods of inductively coupled plasma on the spectrometer OPTIMA-7,000 DV (Perkin Elmer).

According to research years average data on the total composition, it was found that the topsoil contained, mg/kg: As—16.95 ± 0.18; Ca—456.00 ± 18.66; Cd—6.83 ± 0.00; Co—16.85 ± 0.00; Cr—16.43 ± 0.00; Cu—0.99 ± 0.19; Fe—301.73 ± 21.62; Mg—288.93 ± 29.66; Mn—96.65 ± 1.97; Mo—16.23 ± 0.00; Ni—17.04 ± 0.64; Pb—25.71 ± 1.64; Sr—1.89 ± 0.12; Zn—21.72 ± 1.28.

The reference points, such as Clarke for chemical elements, are necessary for the detection of natural soil-geochemical changes and correct assessment of the soil composition of elements, as well as hygienic standards: maximum and approximate permissible concentration (MPC) and APC. The study used the conventional world soil Clarke, proposed by D.P. Malyuga [12].

As a result of determination of chemical substance concentrations (C_c) in soil, it was found that chemical elements form two groups in background sample compared to conventional world Clarke [6]. The excess over the MPC is not revealed.

The elements of the first group—Co, Mo, Pb, Cd are characterized by an increased content in the soil relative to Clarke, which indicates the accumulation of these substances, but not higher than the MAC level. The coefficient of the chemical substance concentration (Cs) is from 1.61 units Co to 3.42 units Cd. The increased content of these elements may be due to geochemical characteristics of the parent rock materials.

More numerous second group of elements—Mn, Cr, Sr, As, Ni, Cu, Zn, Mg, Ca, Fe—shows the deficit relative to Clarke, its critically low content in the soil, suggesting that there is an element subtraction.

Sample preparation for the determination of chemical elements stationary forms, associated with the various soil components was performed to obtain the acid extracts in the microwave decomposition system under pressure Speedwave MWS-2 (made in Germany BERGHOF Products+Instruments Gmb H), using the individually selected mode. This process applies extra pure HNO_3 and HCl acids, which additionally underwent the distillation in the purification system BSB-939-IR.

According to the content, in the form of stationary compounds, bounded with soil components, elements can be conventionally divided into two groups: calcium, iron, magnesium, manganese occur in a large amount in the stationary forms (from 64.98 mg/kg—Mn, to 284.90 mg/kg—Ca) and form a small cluster. Other chemical elements are poor in the soil (0.39 mg/kg—Mo–6.74 mg/kg—Pb) in stationary forms and form a large group of trace elements.

Mobile element forms were determined by the chemical fractionation. In this study, after the comparative overview of the most commonly used chemical elements fractions and extraction agents, the Sposito method appeared to be effective for their recovery. However, some extraction agents of the method were replaced by those with similar properties and device-oriented. Furthermore, the oxide fraction, extracted by the Pampura method, was added to the factions set, offered by Sposito. This fractionation allowed detection of the mobile element compounds in soil samples of 2009–2011, to determine their chemical fractions and ratio, to conduct a statistical analysis of samples.

Three fractions (of the total amount of mobile forms) were found to be dominant for As, Ca, Cd, Co, Mo, Ni, Pb: exchange fraction—26.7 percent (Ni)—29.1 percent (Co), organic—26.6 percent (As)—34.3 percent (Ca) and soluble—18.2 percent (Ca)—21.8 percent (Co). Elements are bounded with both the various soil components (mineral components, hydroxides and oxides, colloids) and the organic substance, in almost equal proportions, form stable complexes with them and play the principal role in plant nutrition. The water-soluble forms are most mobile, potentially more mobile, because they are transported by surface and ground waters, are easily involved in biogeochemical migration, and are available to plants. In addition, the exchange and water-soluble fractions are the reserve of plants nutrition.

Most of the Mg and Sr content in the soil also fall at the organic (28.8 and 28.7%) and the exchange fraction (22.1 and 26.9%); 17.1 percent of magnesium of the mobile forms total content is contained in residual compounds. This is a strategic reserve, bounded with carbonates and bicarbonates Fe, Mn, Al. The manganese proportion in the organic fraction is 60.6 percent, and only 16.7 percent is in the exchange fraction. The iron is in the organic—39.1 percent, oxide—28.6 percent, residual fraction—26.2

percent; it is presented in the soil (mainly) in the form of stable complexes with organic substance. Its occurrence is considerably less in the form of cations and sesquioxides hydrates in colloid-soluble form. Copper is bounded with carbonates and bicarbonates; it is found only in the carbonate fraction (100%). The proportion of zinc oxide fraction is 91.3 percent, the residual fraction is 8.7 percent. This element forms stable surface complexes and is released with the destruction of Fe and Mn hydroxides.

8.4   CONCLUSIONS

The studied morphological parameters, physical and chemical indices, and composition of elements are the key indicators of certain type of soil characteristics, and they determine the fertility and optimality of crop-growing conditions in the areas with a high cryogenic load. They allow the estimation of the soil productive balance, the processes occurring in a complex dynamic system of life, as well as the probability (extent) of pollution under the certain human impact.

The results of the soil characteristic basic indices study can be used in resolving the issues of management and sustainable utilization of soil resources, taking into account the regional features of climatic factors.

KEYWORDS

- **Bulk composition**
- **Chemical elements**
- **Fractionation**
- **Meteorological factors**
- **Soil**
- **The concentration of chemical**

REFERENCES

1. Golov, G. V.; Soil and Ecology Agrophytocenosis Zeya-Bureya Plain. Vladivostok: Dalnauca (*Far-Eastern Science*); **2001,** 162 p. (in Russian).
2. Bome, N. A.; Bome, A. J.; and Belozerova, A. A.; Sustainability of Cultural Plants to Unfavorable Factors of Environment. Tyumen: Publishing House of the Tyumen State University; **2007,** 192 p. (in Russian).
3. Kiryushin V. I. Assessment of Land Quality and Soil Fertility for planning farming Systems and Agrotechnologies. Eurasian Soil Science; **2007,** *40 (7)*, 785–791 p. (in Russian).
4. Perelomov, L. V.; Mn, Pb and Zn Compounds in Gray Forest Soils of the Central Russian Upland. to add: Perelomov, L. V.; and Pinsky, J. L.Eurasian Soil Science, 2003, 36 (6), 610–618 p. (in Russian).
5. Bolshakov, V. A.; Trace Elements and Heavy Metals in Soils. Eurasian Soil Science; **2002,** *7,* 844–849 p. (in Russian).
6. Ozyorskiy, A. Y.; Fundamentals of Environmental Geochemistry: Textbook. Manual Krasnoyarsk: IPK SFU (Regional institute of in-plant training of Siberian federal university). **2008,** 316 p. (in Russian).
7. Shumanova, A. A.; Agroclimatic Resources of the Tyumen Region. Shumanova, A. A.; Sokolov, B. S.; Chulkov M P., et al.; hereinafter referred: Leningrad: Gidrometeoizdat (Publishing House of the State Committee of the USSR on Hydrometeorology and Control of Natural Environment). **1972,** 150 p.

8. Lihenko, I. E.; Selection of spring soft wheat for the conditions of the Northern Urals. Lihenko, I. E.; Omsk: Dissertation of Doctor of Agricultural Sciences; **2004,** 14–19 p. (in Russian).
9. Elnikov, I. I.; Biryukova, O. A.; and Kryschenko, V. S.; The multi-diagnostic value for the prediction of winter wheat yield and optimal content of available phosphorus in calcareous chernozem. *Agrochem.* **2009,** *11,* 7–15 (in Russian).
10. Darmaeva, N. N.; Haydapova, D. D.; Badmaev, N. B.; and Nimaeva, O. D.; Agro-chemical and physico-mechanical properties of frozen soils, which determine their potential sustainability in agricultural use. *Agrochem.* **2009,** *11,* 16–21 (in Russian).
11. Long-cultivated sod-podzolic and podzolic soils, their classification and description. http://big-archive.ru/geography/pedology/46.php (in Russian).
12. Malyuga, D. P.; Biogeochemical Method Explorations for Mineral Deposits. Leningrad: Academy of Sciences of the USSR; **1963,** 264 p. (in Russian).

CHAPTER 9

THE SUSCEPTIBILITY OF ENVIROMENTAL FACTORS ON SEED GERMINATION CHARACTERISTICS AND BIOLOGICAL RESISTANCE OF PLANTS HYBRIDS *TRITICUM AESTIVUM* L.

ELENA I. RIPBERGER, NINA AN. BOME, and ALEXANDER YA. BOME

CONTENTS

9.1   INTRODUCTION

The territory of Northern Zauralye (Tyumen area, Russia) characterized by severe continental climate, sudden change in air temperature, prolonged spring drought, a wide variety of soil types, and other factors limiting the growth and development of crop plants by biotic and abiotic factors. In view of the above-mentioned factors, selection and creation of varieties of high adaptive and productive properties is required.

The purpose of the experiment—the study of the impact of environmental factors on seed germination, seedling formation, and biological stability of the hybrid forms of soft spring wheat.

Spring wheat hybrids were obtained in 2009 by crossing five samples according to incomplete diallel scheme. Parental forms included in the experiment were presented varieties of Tyumen selection (Skent 1 and Skent 3), as well as varieties of foreign selection Lyutescens 70 (Kazakhstan), Hybrid (k-47641, Mexico), and Cara (k-64381, Mexico). Local varieties were treated to a variety *lutescens (Alef.) Manf.*, Mexican varieties selection—to species *eritrospermum* Korn. and *ferrugeneum* (Alef.) Mansf. The starting material is picked up on the results of a comprehensive assessment collection fund of soft spring wheat on the basis of the Tyumen science point of Research Institute of Plant Industry by N.I. Vavilov. Castration and pollination of mother plants were carried out by the method described by V.F. Dorofeyev and coauthor [1].

A field test was conducted on the experimental hybrids Biological Station area "Lake Kuchak" by Tyumen State University in 2010–2013. In an experiment in 2010, we used the method of an individual assessment of each plant. Sowing was carried out with the inclusion of parental units and hybrid forms with a feeding area for each plant 10×20 cm analyzed in 203 plants. In 2011–2013, hybrids of the second, third, and fourth generations (F_2, F_3, F_4) were evaluated families (family—descendants of one plant). Area plots depended on the number of seeds. The number of families studied was 91. Sowing was held on May 6 in 2010; May 10, 2011; May 9, 2012; and May 9—in 2013.

It is known that during the growing period, the plants are exposed to various environmental factors, such as extreme temperatures, lack or excess moisture, nutrient deficiency, disease damage, and other factors that may lead to delay in development and growth, and ultimately to the death of a phase of the plant. The manifestation of many selection-important traits at the phenotypic level, under the circumstances, according to V.P. Shamanin, depends primarily on the rate of the reaction of the genotype to changing environmental conditions [2].

Studies of B.A. Rubin revealed that the hereditary nature of the plant organism are determined by features such as the late- and early maturation, the overall productivity, number of shoots, and resistance to adverse effects of biotic and abiotic factors. But at the same time, the realization of the genetic potential of plants largely depends on the environmental conditions in which they are being development [3].

For the conversion of the resting grain seedling to the optimal combination of environmental need some factors—water, heat, and oxygen—are required. Completeness of germination depends on many factors, including the conditions for the formation and maturation of seeds, storage, heat, and moisture during germination. The major requirement for the seeds to germinate last property is considered, giving strong shoots that cannot only survive in a range of adverse environmental factors, but also thrive and develop. An important task is to preserve the formative processes in the period since the germination of seeds and plants to complete the transition to auto-trophic type of food [4,5]. At the same time, as noted by A.I. Nosatovsky, lack of soil moisture during grain germination and seedling emergence leads to a thinning or delay in germination [6].

In our study, the original and hybrid (F_1; F_2; F_3; F_4) forms of spring wheat were evaluated by field germination and survival of plants in various hydrothermal regime on growing periods. A characteristic feature of the period of the study can be considered as the uneven distribution of rainfall. With the passage of separate development phases of plant spring wheat, moisture deficit was observed against the background of increased air temperatures, which is most clearly evident in 2012. In terms of quantitative characteristics (maximum air temperature, dry periods), in this year's record values observed over the multiyear averages.

In the process of seed germination, there are three phases: physical (seeds absorb water and swell), biochemical (conversion of reserve substances insoluble to soluble), and morphology (the beginning of the growth of the fetus) [7]. Consequently, heat and moisture supply during this period may be crucial for the occurrence of the initial stages of ontogeny.

According to our data when the conditional distribution of the initial forms and hybrid combinations on field germination of seeds into four groups found that during the years of the study, most of the studied samples of spring wheat were characterized by high and very high levels. In 2013, 60.0 percent of the forms were assigned to a group with average field germination (Figure 9.1).

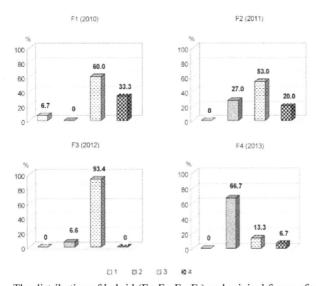

FIGURE 9.1 The distribution of hybrid (F_1; F_2; F_3; F_4) and original forms of spring wheat in the group on field germination of seeds %.
Groups:1—low (**<50%**), 2—moderate (51–70%), 3—high (71–90%), 4—very high (> 90%).

However, the comparative analysis of the starting material and the three generations of hybrids revealed some differences in the number of full shoots.

Thus, the germination of seeds in May 2010 took place at a mean air temperature
of 13.0°C (average long-term value 10.6°C). Although the amount of rainfall in the
first and second decades of the month was minimal (1.0 and 0.4 mm, respectively),
seedling emergence after seeding was amicable. Given that for the formation of seedling, a higher temperature than the seed germination, it can be said that the temperature
regime was favorable. Moisture for the swelling and the formation of sprouts seeds of
spring used inventory.

When comparing parental and hybrid forms, it was found that there was a group
in the F_1 hybrids (10.0%) with low field germination of seeds, which can be associated with a lower supply of nutrients in the small seeds (Figure 9.2). The inhibition of
growth can be caused by trauma to the underdevelopment and endosperm of hybrid
kernels, which is a supplier of nutrients, the fetus.

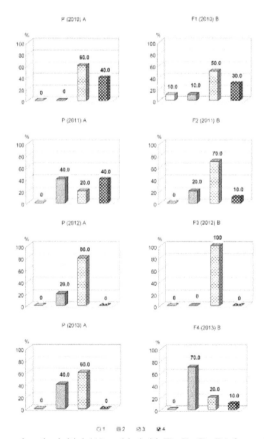

FIGURE 9.2 Comparing the initial (A) and hybrid (F_1; F_2; F_3; F_4) forms (B) of spring wheat
field germination, 2010–2013.

Groups: 1—low (< 50%), 2—moderate (1–70%), 3—high (71–90%), 4—very high (> 90%).

In 2011, the swelling and germination of seeds of parental and hybrid (F_2) forms took place with the lack of moisture in the soil, as the amount of rainfall for May in relation to the long-term average value was only 84 percent, while the average daily air temperature was above normal at 1.3°C.

The distribution of initial varieties and hybrids into groups with different indices of trait revealed that the proportion of samples with high and very high values decreased in comparison with 2011. Most samples (53%) belonged to a group of high field germination of seeds. When compared with the parental and hybrid (F_2) forms on this index revealed that 70 percent of the F_2 hybrids were characterized by a high and 10 percent a very high field germination of seeds (Figure 9.2). At initial varieties, the amount of these groups accounted for 60 percent.

May 2012 was characterized by high air temperature (at 2.4°C above normal) and a rainfall deficit (33.9% of normal). Total precipitation for the decades of the month ranged from 0.3 to 4.0 mm and the number of days with rain was 8. Seed germination and seedling formation took place with warm weather and low precipitation.

Germination of hybrids (F3) varied from 72.8 to 85.0 percent, and on the basis of these values was assigned to one group, characterized by high values of the trait. The initial sign of varying grades are in the range from 62.0 to 82.0 percent. Identification of two groups of samples, the average (20%) and high (80%) of germination, indicating differences in the flow of metabolic processes in the germinating bulked less pronounced adaptation properties of the germ in the transition from mesotrophic to autotrophic nutrition (Figure 9.2).

In 2013, the conditions for seed germination and seedling formation can be considered inadequate by favorable temperature conditions, as the average temperature in May at 1.6°C was below normal. In the first and second decades of the month marked negative temperature: up—1.5°C and 1.9°C, respectively. Reduced air temperature accompanied by fairly frequent precipitation, which together accounted for 140.1 percent last month compared to the average long-term value. In the first decade, the rains occurred each day (from 0.4 to 14.0 mm) and in the second and third decades the precipitation days were lower (4 and 3 days, respectively).

Hybrids F4 under the circumstances, in 2013, 10 percent of the forms were characterized by very high field germination of seeds. The remaining 90 percent of hybrids were assigned to groups with medium and high values of this index (Figure 9.2).

The survival rates of plants during the growing season depend on the genetic characteristics of varieties, and from some combination of environmental factors (soil moisture, air temperature, damage plants by phytopathogenic fungi, etc.).

The average air temperature during the growing season in 2010 was higher than the long-term average. Especially, this value exceeded in May (+2.4°C) and August (+2.9°C). The amount of precipitation in relation to the long-term average in July was 62.6 percent, and in August—77.4 percent. The ontogeny of plants defeat disease, and pest damage, as reflected in the indicators of biological stability. In 73 percent of the studied material are low and the median survival of plants (Figure 9.3).

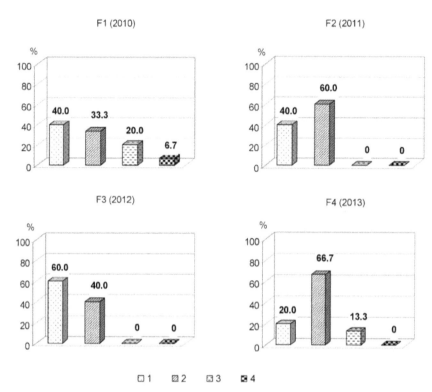

FIGURE 9.3 The distribution of hybrid (F_1; F_2; F_3; F_4) and original forms in biological sustainability, %.

Groups: 1—low (< 50%), 2—moderate (51–70%), 3—high (71–90%), 4—very high (> 90%)

F_1 hybrids compared with their parents were characterized by higher biological stability of plants, as evidenced by the appearance of the fourth group, comprising 10 percent of the forms (Figure 9.4).

The vegetation period of 2011 as a whole (except for the month of July) was characterized by high air temperature. Lack of moisture was recorded in the months of July (67.1% of normal rainfall), August (47.7%), and September (87.0%). The excess amount of precipitation (52.0 mm, which is 49% above normal) during the second half of June contributed to the creation of favorable conditions for the development of pathogenic fungi.

Under these conditions, the distribution of samples of spring wheat in the group-identified general rule is manifested in the same proportion of groups in the analysis, as the whole set of samples, and individually F_2 hybrids and parental species (Figures 9.3 and 9.4). In 40 percent of the cases, we detected the low ability of plants to withstand the effects of adverse environmental factors (Figures 9.3 and 9.4).

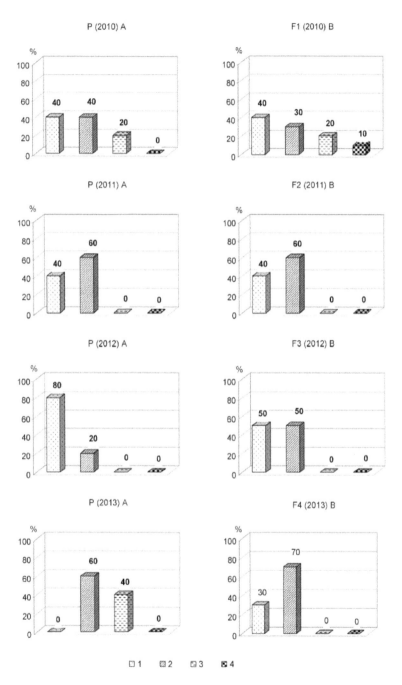

FIGURE 9.4 Comparison of original (A) and hybrid forms (F_1; F_2; F_3; F_4) (B) of spring wheat on biological sustainability of plants during the growing season, 2010–2013.

Groups: 1—low (<50%), 2—moderate (51–70%), 3—high (71–90%), 4—very high (>90%)

Conditions for completion of all stages of ontogeny in 2012 can be characterized as extremely dry. In June, the average daily air temperature is higher than average long-term value for 4.1°C, the amount of precipitation from the norm was 63.7 percent, the uneven distribution of the decades. To stress on the plants during this period can be attributed to significant difference in temperature between 7.6°C (08.06.) and 32.3°C (21.06.). The average daily temperature for the month of July was at 21.4°C for the18.6°C long-term average, the amount of rainfall—24.3 mm at a rate of 84 mm, and the number of days with precipitation—8. Conditions for grain filling and ripening in August had not fared well. Despite the fact that in the second decade of precipitation was 166.5% of the norm, they were not able to eliminate the effects of drought. The average daily temperature for the month is high 17.7°C (14.5°C rate) and the amount of rainfall is 29.3 mm (50.5% of normal).

In general, the studied material dominated the group with low survival spring wheat plants (Figure 9.3). Increased resistance to stress factors were characterized by hybrid F_3, in which there was an equal ratio of groups with low and high survival rate, while 80 percent of the parental varieties have experienced significant inhibition of growth processes and depression.

In 2013, the average daily temperature in June and July was close to the average of many years, while August was warmer than usual to 1.3°C. The growing season characterized by uneven distribution of rainfall resulted in a significant shortage in June and August (about 60% of normal) and abundance in July (142.4% of normal). It should be noted that the excess moisture in July on background sufficiently high temperatures promoted the development of phytopathogenic fungi. The plants of spring wheat marked the defeat of brown leaf rust and patchiness of different etiologies. In a number of cases, the lesion was significant. Integrated display of negative factors is reflected in the survival rates of plants.

In the allocation, the parental forms and hybrids F_4 identified two groups in the number of plants that have passed all phenological phases and formed a full grain. Averages of biological stability possessed 60 percent of the original varieties and hybrids 70 percent. Additional hybrids identified a combination with high survival plants, while parental varieties were 40 percent (Figure 9.4).

Seed germination and survival of plants during the growing season can be considered as two important and inter-related traits, for the ability of seeds to germinate under the circumstances, and integrated plant resistance to environmental factors. In this regard, of particular value forms are capable of forming full-shoots with the further passage of the stages of ontogeny. According to our data, in difficult conditions in 2010, for the integrated display of signs stood hybrid combination ♀ Lyutestens 70 × ♂ Skent 1, which received 100 percent of seedlings and plants, were no deaths. In 2013, this hybrid combination was characterized by the same very high levels of field germination (90.4%), the biological stability (63.5%). According to the 2011 best results were obtained in the two hybrid forms: ♀ Lyutecens 70 × ♂ Skent 3 (germination—83.5%, the survival rate—65.3%) and ♀ Hybrid × ♂ Skent 1 (germination—79.5%, the survival rate—68.1%). In 2012, the best results were obtained in a hybrid combination ♀ Skent 3 × ♂ Skent 1 (germination—84.0%, the survival rate—55%). The average data for the 4-year study of three hybrid combinations stood out: ♀ Lyutescens 70 × ♂ Skent 1, ♀ Hybrid × ♂ Cara, and ♀ Skent 3 × ♂ Skent 1.

9.2 CONCLUSION

Identifying the genetic potential of soft spring wheat and the nature of adaptive responses to changing environmental factors can be used as a proxy field germination of seeds and the biological stability of the plants during the growing season. It is established that the adaptive properties of the seeds during germination and plant development in ontogenesis depends on the genotype and meteorological factors. This shows that the sources of spring wheat varieties are more sensitive to the limiting factors of the environment than hybrids, derived from crosses of these varieties.

KEYWORDS

- **Biological resistance**
- **Environmental**
- **Germination**
- **Hybrid**
- **Triticum aestivum L.**
- **Tyumen area**

REFERENCES

1. Dorofeyev, V. F.; Flowering, Pollination and Hybridization of Plants. Dorofeyev, V.F.; Laptev, J. P.; Tcherkashin, I.; Moscow: "Agropromizdat"; **1990,** 144 p (in Russian).
2. Shamanin, V. P.; Overall Selection and Varieties Knowing Field Crops: Studies. Manual. Shamanin, V. P.; and Truschenko, A. U.; Omsk: Publishing House of the HPE OmGAU; **2006,** 400 p (in Russian).
3. Rubin, B. A.; Course of Plant Physiology. 4th edition. Ed. Rubin, B. A.; Moscow: "Higher School"; **1976,** 576 p (in Russian).
4. Bome, N. A.; Sustainability of Crop Plants to Environmental Stress. Bome, N. A.; Bome, A.Y.; and Belozerova, A. A.; Tyumen: Tyumen State University Publishing House; **2007,** 192 p (in Russian).
5. Laman, N. A.; Ecological validity of the management of production processes in agrophytocenosis. *Ecol.* **1996,** *1,* 10–16 (in Russian).
6. Nosatovsky, A. I.; and Wheat, A. I.; Nosatovsky. Moscow: State Publishing House of Agricultural Literature; **1950,** 407 p (in Russian).
7. Vasko, V. T.; Fundamentals of Seed of Field Crops. Vasko, V. T.; St. Petersburg: "Publishing Lan"; **2012,** 304 p (in Russian).

CHAPTER 10

RELATIONSHIP BETWEEN SOYBEAN VARIETIES, SPAD502 CHLOROPHYLL METER READINGS, AND RHIZOBIA INOCULATION IN WESTERN SIBERIA

INSA KUEHLING, NINA BOME, and DIETER TRAUTZ

CONTENTS

10.1   INTRODUCTION

Due to climate change and an increasing demand for food and fodder, the production of soybeans (*Glycine max*) in Russia as well as in Western Siberia increases [1], especially since the last 5 years (Figure 10.1).

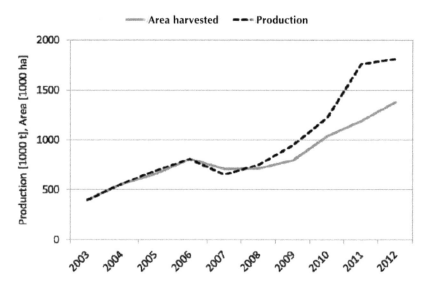

FIGURE 10.1   Production and harvested area of soybeans in Russia [2].

Soybeans as a legume crop species do have a close relationship between the availability of nitrogen either from N_2-fixation or soil N and the net leaf photosynthetic rate [3]. Biological nitrogen fixation requires necessarily the inoculation with specific rhizobia bacteria, for example, *Bradyrhizobium japonicum* [4]. Soils usually lack *B. japonicum* strains [5]. Therefore, it is important to inoculate seeds especially if soy is planted on the field for the first time, otherwise no nodulation will occur [6]. High variability in nodulation [7] and N_2 fixation [8] between soybean genotypes is reported. By comparing the leaf chlorophyll content of plants with and without nodules, it is possible to quantify how much inoculation may increase the N-uptake [3]. Leaf chlorophyll metering, for example, by Minolta SPAD-502, appears to be a useful technique for determining the nitrogen status of soybeans [9].

Within the framework of the interdisciplinary German-Russian project SASCHA [10], field trial was conducted to investigate the relationship between soybean varieties, rhizobia inoculation, and SPAD-502 chlorophyll meter readings.

10.2   PROJECT BACKGROUND

The province Tyumen in Western Siberia is of global significance in terms of agricultural production, carbon sequestration, and biodiversity preservation. At the same time, the region is the heart of Russia's flourishing gas and oil production and has been the scene of recent large-scale agricultural change [10]. Abandonment of arable land and changes in the use of permanent grasslands were triggered by the dissolution of the Soviet Union in 1991 and the following collapse of the state farm system. The peatlands, forests, and steppe soils of Western Siberia are one of the most important carbon sinks worldwide [11]. These carbon stocks are, if deteriorated, an important source of radioactive forcing even in comparison to anthropogenic emissions. This situation is aggravated by recent and future developments in agricultural land use in the southern part of Western Siberia. The increase of drought risk in the steppe zone of Northern Kazakhstan and Southern Siberia will trigger a northward shift of the West Siberian grain belt into the forest steppe and pre-taiga zone. In conjunction with climate change, agricultural expansion, and intensification in these regions will increase the threat of large-scale release of greenhouse gases from reclaimed peat areas.

SASCHA aims to provide management practices to cope with these far-reaching changes. In particular, as farm-scale strategies are being developed for increased efficiencies in crop production systems.

10.3   MATERIALS AND METHODOLOGY

A field trial with soybean varieties from Germany and Siberia was established at the experimental site "Lake Kuchak" of Tyumen State University in Western Siberia (57°20'56"N, 66°3'24"E) in May 2013 (Figure 10.2).

FIGURE 10.2   Experimental station "Lake Kuchak"/Tyumen State University in Tyumen/ Western Siberia.

Kuchak shows continental climate with a mean min/max temperature of −2.9/6.9 °C (range from −42 to 38 °C) and yearly average precipitation of 450 mm. The experimental site (podsolic soil) was irrigated during the trials as required.

Two early/very early varieties from Germany (Augusta, Aveline) and one regional Siberian (Sibniik315) were grown in completely randomized block design with four repetitions in two variants: with and without *B. japonicum* inoculation. The treatment was done directly before seeding (May 15, 2013) with the peat-based two components product "Force48."

Chlorophyll content was measured at the last full-developed leaf using a Minolta SPAD-502 device as an average value from 30 measurements per plot. The relationship between SPAD values and chlorophyll content is nonlinear for soybeans and follows an exponential equation [12]. SPAD meter readings as well as leaf chlorophyll content is closely related to leaf nitrogen content [13].

Numbers of nodules were counted on carefully digged out roots at randomly collected plants from each plot.

10.4   RESULTS AND DISCUSSION

The number of nodules has shown significant differences between inoculated and control among all varieties. No plant of the control group was infected by bacteria, but at every inoculated plant nodules developed (Figure 10.3). At roots from the regional Siberian variety, the highest number of nodules was counted. End of July (full flowering) soy plants from Aveline had from 9 to 14 nodules, those from Augusta between 4 and 18 while Sibniik315 had 9–40.

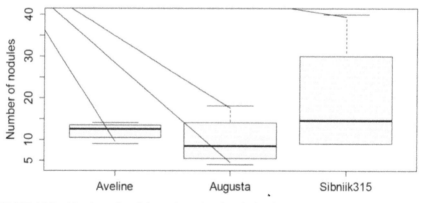

FIGURE 10.3   Number of nodules at inoculated varieties from Germany (Aveline, Augusta) and Siberia (Sibniik315) at full flowering (BBCH 65) on July 20, 2013.

At all plants, the nodules were located close to the main root with diameters between 3 and 5 mm at that stage of development (Figure 10.4(B)). The color of the nodules was light red in the inside, which means that they were slightly active (Figure 10.4(C)).

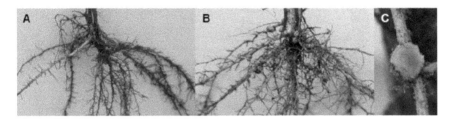

FIGURE 10.4   Roots from two Sibniik315 plants without (A) and with (B) inoculation, sliced nodule (C) during full flowering (BBCH 65).

According to the amount of nodules among the inoculated variants also the SPAD-502 chlorophyll meter readings showed differences. SPAD values tended to be higher with inoculation than in the untreated control group (Figures 10.5–10.7). Significant differences appeared at the third measurement (latest observed stage) during the middle of seed development. In average, Aveline showed the highest increase in SPAD values (+20.3%, Figure 10.5) compared to control, followed by Augusta (+12.7%, Figure 10.6) and Sibniik315 (+9.8%, Figure 10.7).

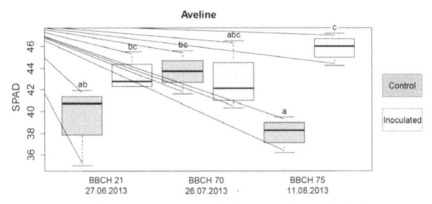

FIGURE 10.5   SPAD meter readings from Aveline at beginning of side shoot formation (BBCH 21), end of flowering (BBCH 70) and mid of seed development (BBCH 75).

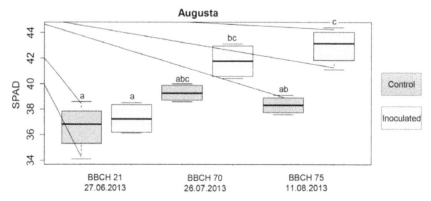

FIGURE 10.6 SPAD meter readings from Augusta at beginning of side shoot formation (BBCH 21), end of flowering (BBCH 70) and mid of seed development (BBCH 75).

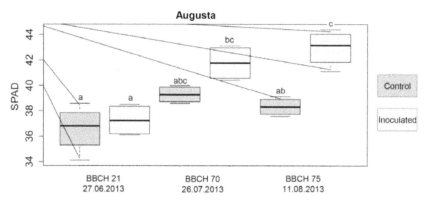

FIGURE 10.7 SPAD meter readings from Sibniik315 at beginning of side shoot formation (BBCH 21), end of flowering (BBCH 70) and mid of seed development (BBCH 75).

Comparable with the result reported by [8], the development of leaf chlorophyll content was specific for every variety. Among inoculated plants, no significant differences at all three observations could be measured in Aveline (Figure 10.5, clear). In contrast, SPAD values of inoculated Augusta and Sibniik315 increased significantly with time (Figures 10.6–10.7: BBCH 21 to 70), but no significant differences occurred without treatment (Figures 10.6–10.7 gray).

The highest Sibniik315 SPAD value was noticed at the second measurement (BBCH 70, inoculated), which decreased slightly in BBCH 75 (Figure 10.7, clear) as well as in the control. This may imply a better adaption of the Siberian variety to the local climate condition (short vegetation period) than the German breeding because relocation effects from vegetative to generative plant organs occurred earlier. Nevertheless, inoculated Sibniik315 showed the highest number of nodules; SPAD measurements were less distinctive (not significant) and higher compared to control.

In addition to the short vegetation period in Siberia, also soil temperatures play an important role for plant physiological processes of nodulating legumes.

TABLE 10.1   Soil temperature in 12–20 cm depth. Average of 3 measurements ± standard derivation

Date (dd/mm/yy)	BBCH	Soil temperature (°C)
12/05/13	before seeding	10.0 ± 1.0
28/0513	09	14.6 ± 1.1
19/06/13	13	21.3 ± 2.5
08/08/13	74	20.0 ± 1.0

Optimal conditions for symbiotic fixation are soil temperatures from 25 to 30°C [14–16]. In contrast, the measured soil temperatures in Kuchak were only ±20°C until BBCH 74 (Table 10.1) and may cause the slow increase of symbiotic activity.

10.5   CONCLUSION

Inoculation of soybeans with *B. japonicum* is necessary, otherwise no nodules for atmospheric nitrogen fixation will be developed under Siberian conditions. Due to low soil temperatures below the optimal range for symbiotic activity, the significant effect of inoculation shows up at a late stage of development (BBCH 75). Therefore, the advantages of inoculation may be limited by Siberian climate conditions.

Local varieties are better adapted to local conditions, but genotypes from other regions must be used to improve the local breeding.

KEYWORDS

- **Chlorophyll**
- **Inoculation**
- **Nitrogen fixation**
- **Nodulation**
- **Soybean**

REFERENCES

1. Kiselev, S.; Romashkin, R.; Nelson, G. C.; Mason-D'Croz, D.; and Palazzo, A.; Russia's Food Security and Climate Change: Looking into the Future. Economics—The Open-Access, Open Assessment E-Journal; **2013,** 7 (in Russian).
2. Faostat Crops—Russia—Soybeans—2009–2005—Production Quantity. Food and Agriculture Organization of the United Nations, **2011.** http://faostat.fao.org/site/567/default.aspx#ancor Accessed on 20/08/2012.

3. Vollmann, J.; Walter, H.; Sato, T.; and Schweiger, P.; Digital Image Analysis and Chlorophyll Metering for Phenotyping the Effects of Nodulation in Soybean. Computers and Electronics in Agriculture; **2011,** *75,* 190–195.

4. Werner, D. Biologische Stickstoff-fixierung (BNF). In Keller, E.; Hanus und K-U, H.; Heyland Handbuch des Pflanzenbaues 3 Knollen- und Wurzelfrüchte, Körner und Futterleguminosen. Stuttgart: Ulmer; **1999.**

5. Hiltbold, A. E.; Patterson, R. M.; and Reed, R. B.; Soil populations of rhizobium japonicum in a cotton-corn-soybean rotation. *Soil Sci. Soc. Am. J.* **1985,** *49,* 343–348.

6. Solomon, T.; Pant, L. M.; and Angaw, T.; Effects of Inoculation by Bradyrhizobium japonicum Strains on Nodulation, Nitrogen Fixation, and Yield of Soybean (Glycine max L. Merill) Varieties on Nitisols of Bako. Western Ethiopia: ISRN Agronomy; **2012,** 1–8.

7. Cregan, P. B.; Keyser, H. H.; and Sadowsky, M. J.; Host Plant Effects on Nodulation and Competitiveness of the Bradyrhizobium Japonicum Serotype Strains Constituting Serocluster 123. *Appl. Environ. Microbiol.* **1989,** *55,* 2532–2536.

8. Herridge, D.; and Rose, I.; Breeding for Enhanced Nitrogen Fixation in Crop Legumes. Field Crops Research; **2000,** *65,* 229–248.

9. Gwata, E. T.; Wofford, D. S.; Pfahler, P. L.; and Boote, K. J.; Genetics of promiscuous nodulation in soybean: nodule dry weight and leaf color score. *J. Hered.* **2004,** *95,* 154–157.

10. SASCHA Nachhaltiges Landmanagement und Anpassungsstrategien an den Klimawandel für den Westsibirischen Getreidegürtel, **2013.** http://www.uni-muenster.de/SASCHA.

11. SASCHA Sustainable Land Management and Adaptation Strategies to Climate Change for the Western Siberian Corn Belt. Progress Report; **2012.**

12. Markwell, J.; Osterman, J. C.; and Mitchell, J. L.; Calibration of the minolta SPAD-502 leaf chlorophyll meter. *Photosynth Res.* **1995,** *46,* 467–472.

13. Fritschi, F. B.; and Ray, J. D.; Soybean leaf nitrogen, chlorophyll content, and chlorophyll a/b ratio. *Photosynth.* **2007,** *45,* 92–98.

14. Lynch, D. H.; and Smith, D. L.; Soybean (Glycine Max) Modulation and N2-Fixation as Affected by Exposure to a Low Root-Zone Temperature. Physiologia Plantarum; **1993,** *88,* 212–220.

15. Lynch, D. H.; and Smith, D. L.; The Effects of Low Root-Zone Temperature Stress on Two Soybean (Glycine Max) Genotypes when Combined with Bradyrhizobium Strains of Varying Geographic Origin. Physiologia Plantarum; **1994,** *90,* 105–113.

16. Zhang, F.; Dashti, N.; Hynes, R. K.; and Smith, D. L.; Plant Growth Promoting Rhizobacteria and Soybean [*Glycine Max*(L.) Merr.] Nodulation and Nitrogen Fixation at Suboptimal Root Zone Temperatures. Annals of Botany; **1996,** *(77),* 453–460.

CHAPTER 11

PROBLEMS OF CARYOPSIS AND STABILITY OF WINTER AND SPRING FORMS OF CEREALS TO PHYTOPATHOGENIC FUNGI OF GENUS *FUSARIUM* LINK

NINA AN. BOME, ALEXANDER YA. BOME,
NATALJA N. KOLOKOLOVA, and YULIA B. TROFIMOVA

CONTENTS

11.1 INTRODUCTION

The problem of stability to diseases and creation of steady cultivars continue to remain actual for selectionists. Laboratory methods of estimation have a considerable importance for the acceleration of plant-breeding work on the selection of steady cultivars. Often they are express methods, because they provide the information in a very short time period, and the estimations of stability are performed at all stages of organogenesis of plants.

Crop-growing conditions in different areas of the Tyumen region are formed unevenly. The climate is affected by cold air masses from the Arctic Ocean and from Siberia, as well as by dry winds blowing from Kazakhstan and Central Asia. The climate is typically continental, and all the climatic factors vary greatly over the years, both in strength and duration. There are elements of the climate similar to the western region (dry summer periods), circumpolar areas (very short and cold growing season), and the deserts of the south (dry, oppressive weather from spring to fall) [1].

Harsh cold winters, relatively short summers, short springs, and autumns, a small frost-free period, and sharp changes in temperature during the year and even during the day characterize agricultural areas of the Tyumen region. One of the causes of the yield decrease in agricultural crops, including cereals, is the growth of infection of the most dangerous diseases. Plants suffer from both pathogens belonging to soil pathological complex (root rot, *Fusarium* wilt, etc.) and air-spread infections (rust, *Septoria*, smut disease, powdery mildew, etc.) [2].

Phytopathogenic fungi of the genus *Fusarium* belong to the most dangerous among more than 350 species of toxigenic fungi known in agriculture [3]. It is shown that the contamination of seeds of spring wheat can occur both in the hidden and explicit form, and to a large extent it is determined by the varietal characteristics [4].

11.2 MATERIAL AND METHODOLOGY

Fungi were singled out from seven genera of pathogenic microflora of grains of spring wheat, barley, and rye cultivars of different eco-geographical origin and different years of harvest. Of these genera, *Alternaria*, *Helminthosporium*, *Trichothecium*, *Tilletia*, and *Fusarium* are representatives of field microflora; *Mucor* and *Penicillium* belong to mold species. The fungal spores of the genus *Alternaria* dominated on most grains of all cultivars. Pathogens from the genera *Helminthosporium* and *Fusarium* of the most harmful type, causing root rot and spot, were detected.

An index characterizing the development of disease was calculated as follows:

$$P_b = \frac{\sum_{i=1}^{k} a_i \cdot i}{N} \cdot 100\%,$$ where i and ai are degree of damage and the corre-

sponding number of ill seedlings, respectively; N is the amount of plants in a test; and k is the maximum degree of damage [5–7]. Degree of damage was set according to the scale of Rusakov and Strachov as follows: 0—no damage; 1—below 10 percent of seedling surface is damaged; 2—11–25 percent; 3—26–50 percent; 4—more than 50 percent of plant surface is damaged.

11.2.1   SELECTION OF FUNGI IN A PURE CULTURE

Seeds were washed in 0.1 percent solution of streptomycin. After a torrefaction, they were placed in a moist chamber at the temperature of 24–26°C. In 2–3 days, seeds with germinating mycelium were put on agar in double-dish (in sterile terms) and kept in thermostat at the temperature of 24–26°C. Pure culture was saved on a slanting agar.

Taking into account the fact that pathogens of the genus *Fusarium* are rather common in the cereals (both spring and winter forms) in Tyumen region and can cause significant yield losses, we conducted laboratory and field studies on the biology of this genus. The experiments included method of phytopathological analysis of seeds with the calculation of the disease index [5–9].

11.3   RESULTS AND DISCUSSION

The incidence of seeds by the pathogenic fungi depends in a great extent on the weather conditions during the ripening and harvesting of grain. Therefore, the formation of the microflora of winter rye grains was studied in the years 2002 (11 samples), characterized by high humidity, and 2003 (17 samples), favorable for ripening grain.

Regardless of the growing conditions, the fungal spores of the genus *Alternaria* dominated in most of grains. Fungi of the genus *Trichothecium* in 2002 were found in 17 percent of seeds, and in 2003 in 1.5 percent of seeds (Figure 11.1).

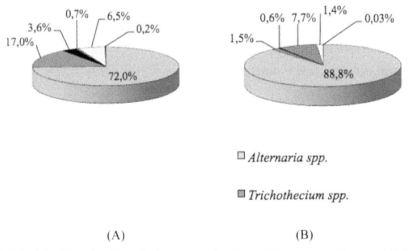

☐ *Alternaria spp.*

▨ *Trichothecium spp.*

(A)   (B)

FIGURE 11.1   The microflora of winter rye grains from different years of harvest: (A) 2002 and (B) 2003.

Among the most harmful pathogens, causing root rot and spot, pathogens from the genera *Helminthosporium* and *Fusarium* were detected. The number of grains affected by Fusarium was 3.6 percent in 2002 and 0.6 percent in 2003; blight affected 0.7 and 7.7 percent of grains, respectively.

Effect of long-term storage of seeds on the composition of the microflora (7 years) is shown on the example of three cultivars of winter rye («Getera», «Pishma», «Voshod1»). Grains of yield 2002 are characterized by greater cultivar of microorganisms (5 genera), with a predominance of *Alternaria* spp. and *Trichothecium* spp. In samples in 1996 was dominated by representatives of the mold. Approximately 100 percent of the seeds were infected with *Mucor* SPP. Long-term storage affected the laboratory germination of seeds, as well as the growth and development of seedlings, causing a significant reduction in the length of roots (54.6%), length of shoots (42.3%), and weight of shoots (46.8%) (Table 11.1).

TABLE 11.1   The impact of long-term storage on the morphometric parameters of seedlings (average of cultivars)

Feature	Control	Storage 7 years·	Deviation from the control (%)
Laboratory germination, %	94.0	34.3	−63.5
Length of roots, mm	134.1 ± 4.00	60.9 ± 3.92*	−54.6
Number of roots, pcs.	5.5 ± 0.08	5.0 ± 0.69	−9.1
Weight of the roots, g	4.9	4.8	−2.0
The length of coleoptile, mm	47.4 ± 0.85	32.6 ± 1.93	−31.2
The length of the stem, mm	107.3 ± 2.22	61.9 ± 4.57*	−42.3
Weight of the plant shoots, g	5.26	2.8	−46.8

Note: *—The differences with control were statistically significant at P < 0.05; °control—harvest 2002; and • storage of 7 years—harvest of 1996 year.

It is known that the cultivation of plants in different conditions gives the possibility to study their investigate the occurrence of quantitative characteristics and the resistance to environmental stress including pathogen that allows to detect ecological plasticity of the material under test.

Evaluation of the presence of fungal infection of seeds of cultivar «Chulpan» grown in soil and climatic conditions of northern forest-steppe region of Tyumen (Tyumen) and Middle Ural (Yekaterinburg), revealed a high infestation.

The grains obtained in Yekaterinburg were affected by six genera of pathogenic fungi, of which the most numerous were *Alternaria* spp. (40.0%), *Mucor* spp. (28.7%), and *Trichothecium* spp. (25.4%). Members of the genus *Helminthosporium* spp., *Fusarium* spp., and *Penicillium* spp. were found on 0.3, −1.8, and 3.9 percent of seeds, respectively. The grains grown in the northern forest-steppe of the Tyumen region were infected by *Alternaria* spp. (68.7%), *Trichothecium* spp. (14.9%), and *Fusarium* spp. (8.8%); *Mucor* spp. was found in 7.6 percent of seeds; fungi of the genus *Penicillium* spp. were not detected.

The study revealed differences in morphometric characteristics of seeds. The seeds grown at the experimental plot in the city of Tyumen formed seedlings with more developed primary root system (the length and number of roots) and coleoptile (L) compared with the seeds of Yekaterinburg. No significant differences were found in the length of shoots and the mass of roots and shoots.

In our experiment conducted in the laboratory on four cultivars of soft spring wheat, we observed the dependence of the ability of seeds to normal germination on their contamination by pathogens. Cultivar Tyumenskaya 80 had the lowest laboratory germination of seeds among cultivars—88.2 percent at the maximal level of infection. In less contaminated cultivars (Saratovskaya 57, Comet, Mir 11), seed germination ability was higher:—98.5, 99.0, and 99.3 percent for Mir 11, Comet, and Saratovskaya 57, respectively.

Fungi of *Fusarium* genus are the most common pathogens among soil infections. They cause disease of roots and root collar, which leads to the death of productive stems, and the empty spike of infested plants.

According to the percentage of seedlings affected by the disease, cultivars Comet, Tyumenskaya 80, and Mir 11 were classified as middle susceptible (P_b = 28.15–30.45%, 21.59–28.43%, and 20.40–29.50%, respectively), and cultivar «Saratovskaya 57» as low susceptible (P_b = 12.48–18.48%) (Table 11.2).

TABLE 11.2   Evaluation of spring wheat samples on infectious background for resistance to *Fusarium* spp.

Cultivar	Phytopathological analysis of seeds		Benzimidazole method		The rots of root		Sum of degrees of damage
	P_b (%)	Degree of damage	P_b (%)	Degree of damage	P_b (%)	Degree of damage	
Tyumenskaya 80	21.59	2	24.00	4	25.55	2	8
Comet	28.15	1	35.43	1	21.87	3	5
Saratovskaya 57	18.48	4	34.35	2	21.11	4	10
Mir 11	20.40	3	24.27	3	28.89	1	7

Note: > 40 percent—susceptible, 20–40 percent—middle susceptible, and 10–20 percent—low susceptible.

A stronger root growth and inhibition of vegetative parts were observed in the study of morphometric parameters on the background of the infected seedlings (Table 11.3).

TABLE 11.3   Quantitative characteristics of spring wheat samples on the infectious background of *Fusarium* spp.

Cultivar	Option	Length of sprout $X \pm m_x$ (cm)	Length of roots $X \pm m_x$ (cm)
Comet	Control	23.52 ± 0.61	14.09 ± 0.43
	Experiment	21.75 ± 0.81	$20.39 \pm 0.81^*$
Saratovskaya 57	Control	22.09 ± 0.95	18.53 ± 0.33
	Experiment	20.83 ± 0.71	19.30 ± 0.96
Mir 11	Control	24.25 ± 0.75	19.28 ± 0.71
	Experiment	$19.63 \pm 0.75^*$	20.72 ± 0.62
Tyumenskaya 80	Control	26.02 ± 0.60	18.96 ± 0.57
	Experiment	$19.81 \pm 0.67^*$	$24.96 \pm 0.42^*$

Note: *—The differences were statistically significant at $P < 0.05$.

Cases of stimulation of the growth processes of infected plants are described by Rodigin, M.N. [10]. Often, this phenomenon is temporary and related with the physiological characteristics of the pathogen. Intensive growth of roots and lagging behind of aboveground parts of the plants can probably be explained by the fact that the introduction of the pathogen into the roots of the plants leads to blockage of vascular system, disrupts the transport of water and dissolved substances, reduces the rate of photosynthesis and, therefore, produces a delay of plant development.

Extensive study including the assessment of the intensity of seed germination, seedling variability of quantitative traits and primary root system, have allowed to identify cultivars of spring wheat «Tyumenskaya 80», «Mir 11», «Saratovskaya 57», as the most resistant to infection.

Productivity of winter crop forms is dependent on a number of biotic (pathogens) and abiotic (temperature, rainfall, etc.) factors. Pathogenic fungi that cause disease play a negative role in plant growth and development. In particular, the snow mold, which is caused by *Microdochium nivale* (Fr.) Samuels and I.C. Hallett (*Fusarium nivale* Ces. ex Berl. and Voglino), is dangerous. It is widely specialized facultative parasite, always present in the soil.

One of the factors that determine the development of the fungus is the temperature. *Fusarium nivale* Ces. begins to develop at 5°C; the optimal growth is observed at 11–17°C [11, 12]. In our laboratory studies performed with U.B. Trofimova [13], the effect of temperature on the rate of development of the fungus was studied. By cultivating the fungus in the thermostat at 20°C, 10°C, and 5°C on potato glucose agar in Petri dishes with threefold repetition, we determined the diameter of the colony and especially sporulation.

The lowest rate of growth of the fungus was recorded at 5°C. Beginning of the growth in this variation was observed on the 8th day after sowing. Fungal colonies reached the diameter of Petri dishes on 42th day, with sporulation recorded only on 56th day (Figure 11.2).

At a temperature of 10°C on the 4th day of the experiment, the diameter of the colony was equal to 12.5 mm, and after 28 days Petri dish was completely occupied by the fungus. In this variant sporulation happened much earlier—on the 16th day. The fastest growth of *Fusarium nivale* Ces. colony was observed at 20°C. The active beginning of growth was already evident on the second day, sporulation was observed on the sixth, while on eighth day of the experiment the colony's diameter reached 90 mm.

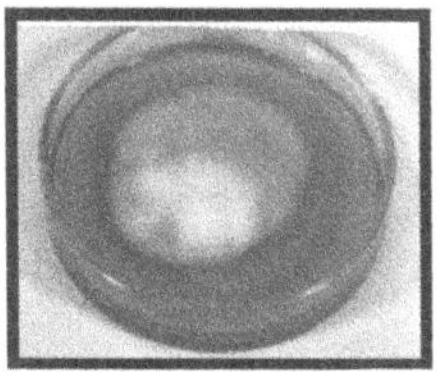	20°C	10°C	5°C
Beginning of the growth	2 days	4 days	8 days
Start sporulation	6 days	16 days	56 days
The diameter of the colony	87.7 ± 2.03 mm	33.0 ± 2.00 mm	90.0 ± 0.00 mm

FIGURE 11.2   Temperature dependence of the development of fungus Fusarium nivale Ces.

Development of snow mold is determined by weather conditions of the spring period and is not observed every year, so any conclusion on plant resistance can only be made in the years of strong manifestation of the disease. One of the conditions to obtain reliable results in the determination of resistance is the creation of artificial conditions ensuring optimal infection load. This background on the experimental site was created by application into soil of an aqueous suspension of spores and mycelia of pure 14-day culture of *Fusarium nivale* Ces. Infectious load was 10^6 conidia/ml of inoculum (500 ml/m² of soil). Infection was carried out in autumn in the phase of bushing out before snow cover. Estimate of snow mold infection of plants was carried out 10 days after snow melting in the early resumption of the growing season according to methodical guidelines of Kobylyansky [14], on a scale developed by Andreev and Molchanova [15] and Geshele [16].

A study of infection in vivo under hard infectious background revealed that harmfulness of snow mold manifested in the reduction of such morphometric characteristics of winter rye as plant height, leaf area, and productivity traits. Decline by more than 50 percent was observed in leaf area per 1 m², number of grains per plant, grain weight per spike and plant. There was a strong development of the disease on the infectious background, which resulted in the decrease of yields compared to the control samples by 38.1 percent on average.

11.4 CONCLUSIONS

It is shown on the example of cultivars of winter rye that in the process of storage is reduced amounts of the seed, infected by the field microflora, and the portion of seeds staggered by a mould microflora increases. After 550 days laboratory germination, length a mass of seedlings decreased by 77, 32, and 37.5 percent, respectively. A decline of these indexes was more considerable at the protracted (7 years) storage.

Microflora of caryopsis, ability of seeds to germinate, and formation of biomass of seedlings depend on meteorological factors. In cool–moist terms, *Alternaria* spp. and *Trichothecium* spp. prevailed in microflora.

Seeds of winter-annual rye (a cultivar Chulpan), obtained in different ecological terms: Tyumen and Ekaterinburg regions differed in quality, susceptibility to pathogenic fungi, and the display of morphometric signs of plants in early ontogenesis.

Laboratory methods proved to be effective in the selection of plastic cultivars of spring and winter forms of grain-crops for northern forest-steppe of Tyumen region.

The information obtined on the first stages of development allows to control the stability of plants in later ontogenesis.

In the growing season, characterized by a long warm autumn, conditions were favorable for active growth of the pathogen. Effect of pathogen was aggravated by soil and air drought in spring and summer. In the experimental variant with infectious load, the inhibition of growth processes was observed, which manifested in significant reduction in breeding-valuable features to 26.21–67.70 percent.

Cultivars «Chulpan», «IL men» «Iset», and «Supermalysh 2» belong to the group of resistant cultivars of winter rye, whereas the cultivars «Voshod 1», «8s-191 Rossianka × Getera», «Desnyanka × Imerig», «Tetra», and «Siberia» have medium susceptibility.

KEYWORDS

- **Infection background**
- **Microflora**
- **Pathogen stress**
- **Phytopathogenic fungi**
- **Stability**
- **Weevil**

REFERENCES

1. Ivanov, P. K.; Spring Wheat. Moscow: Publishing House Kolos; **1971,** 328 p (in Russian).
2. Kosogorova, E. A.; Protection of Field and Vegetable Crops from Diseases. Tyumen: Publishing House of the Tyumen State University; **2002,** 244 p (in Russian).

3. Kudayarova, R. R.; Mitotoksiny. Problems and Prospects of the Development of Innovation in Agricultural Production. All-Russian Scientific and Practical Conference of the XVII Specialized Exhibition AgroComplex-2007. Ufa: Bashkir State Agrarian University; **2007**, Part 2. 79 p (in Russian).

4. Khairulin, R. M.; and Kutluberdina, D. R.; The Prevalence of Fungi of the Genus *Fusarium* in Grain of Spring Wheat in the Southern Forest of the Republic Bashkortostan. Khairulin, R. M.; and Kutluberdina, D. R.; Bulletin of the Orenburg State University; **2008**, *12*, 32–36 p (in Russian).

5. Naumova, N. A.; Analysis of Seeds to Fungal and Bacterial Infection. Leningrad: Publishing House Kolos; **1970**, 32 p (in Russian).

6. Evaluation of Crops for Resistance to Diseases in Siberia. Novosibirsk: Guidelines; **1981**, 48 p. (in Russian).

7. Mikhailina, N. I.; Comparative Evaluation of Methods for Determining the Severity of Root Rot of Spring Wheat. Agricultural Biology; **1983**, *4*, 95 p (in Russian).

8. Zrazhevskaya, T. G.; Guidance on the Study of the Stability of the Grass to the Agents of Diseases of the Conditions for Non-Chernozem Zone of the Russian Soviet Federative Socialist Republic. Leningrad: All-Union Institute of Plant Protection; **1977**, 60 p (in Russian).

9. Zrazhevskaya, T. G.; Determination of the Resistance of Wheat to Common Root Rot. Mycology and Phytopathology; **1979**, *13(3)*, 58 p (in Russian).

10. Rodigin, M. N.; General Phytopathology. Moscow: Publishing House High School; **1978**, 365 p (in Russian).

11. Rubin, A.; Crop Physiology IV. Leguminous Plants. Perennial Grasses. Cereals (Rye, Barley, Oats, Millet and Buckwheat). Moscow: Publishing House Moscow State University; **1970**, 654 p (in Russian).

12. Yakovlev, N.; Phytopathology. Programmed Instruction. Moscow: Publishing House Kolos; **1992**, 384 p (in Russian).

13. Trofimova, U. B.; and Bome, N. A.; A Scholarly Theoretical Peer-Reviewed Journal "Plant Protection News". Journal of Plant Protection. Parameters of Snow Mold Damage and Resistance of Winter Rye to Illness. St. Petersburg: Pushkin; **2006**, *1*, 33–36 p (in Russian).

14. Guidelines for the Study of the World Collection of Rye. Ed. Kobylyansky, V. D.; Leningrad: All-Union Institute of Plant Growing of N. I. Vavilov; **1981**, 20 p (in Russian).

15. Andreev, V.; and Molchanov, O.; Snow Mold of Winter Grains (Methods of Study and Control Measures). Moscow: Research Institute of Technic-Economic Researches; **1987**, 46 p (in Russian).

16. Geshele, A. A.; Bases of Phytopatological Estimation are in the Selection of Plants. Kolos: Moscow; **1978**, 206 p (in Russian).

CHAPTER 12

EVALUATION OF COLLECTION OF MEDICINAL HERBS ACCORDING TO PHYTONCIDIC ACTIVITY

ALENA N. MICHAJLOVA, NINA AN. BOME,
SVETLANA G. SHICHOVA, and NATALIA N. KOLOKOLOVA

CONTENTS

12.1 INTRODUCTION

The science discovered the importance of medicinal herbs, their biological value, and peculiarities of their effect on organism and created many medical products, which are successfully applied under different diseases [1].

Later scientists discovered one of the most important peculiarities of phyton-cides—the specificity of their effect [2]. Phytoncidic activity of most of plants is more expressed during spring and summer. The duration of high level of phytoncides production is different for different species; and in some cases, it is rather limited in time [3]. Phytoncidic volatiles and phytoncidic tissue saps can be an obstacle to the multiplication of bacteria and fungus [4].

The aim of this chapter is the research of phytoncidic properties of some species of medicinal herbs, namely (1) to compare herb species according to phytoncidic activity and (2) to determine antibiotic activity of intracellular components of medicinal herbs relatively to cultures of conditionally pathogenic microorganisms.

12.2 MATERIALS AND METHODOLOGY

The study was performed at the cathedra of botany, biotechnology and landscape architecture of Tyumen State University in 2011–2013.

A total of 30 species of medicinal herbs from 14 families were studied, namely the following:

Rosaceae (*Potentila erecta* L., *Filipendula ulmaria* L., *Fragaria vesca* L., *Agrimonia eupatoria* L., *Potentila recta* L.); *Brassicaceae* (*Armoracia rusticana* Lam.); *Boraginaceae* (*Symphytum officinale* L.); *Scrophulariaceae* (*Digitalis grandiflora* Mill); *Geraniaceae* (*Geranium pratense* L.); *Lamiaceae* (*Prunella vulgaris* L., *Origanum vulgare* L., *Nepeta cataria* L., *Leonurus cardiaca* L., *Betonica officinalis* L., *Phlómis tubérosa* L., *Melissa officinalis* L.); *Apiaceae* (*Eryngium planum* L.); *Urticaceae* (*Urtica cannabina* L.); *Papaveraceae* (*Chelidonium majus* L.); *Valerianceae* (*Valeriana officinalis* L.); *Asteráceae* (*Tussilago farfara* L., *Tanacetum vulgare* L., *Calendula officinalis* L. *Cichorium intybus* L., *Echinacea purpurea* Moench.); *Liliáceae* (*Poligonatum officinale* All., *Lilium martagon* L., *Allium fistulosum* L., *Allium schoenoprasum* L.), *Fabaceae* (*Glycyrrhiza uralensis* L.).

The material was taken from the collection of medicinal herbs, growing at the experimental plot of the biological station "The Lake Kuchak" of Tyumen State University situated in the Nizhnetavdinsky district of Tyumen region. The territory is moderately wet, and hydrothermic coefficient is 1.2–1.3. Annual precipitation total is 350–380 mm. The sum of air temperatures above 10°C is 1,700–1,900 degree-days, and the length of the corresponding period is 114–123 days. Droughts of low and moderate intensity are rather frequent [5].

In 2011 and 2012, laboratory studies were conducted using the method of holes in the thickness of agar, aimed at determining the antibiotic activity of medicinal herb extracts relatively to conditionally pathogenic microorganisms: *Staphylococcus aureus, Staphylococcus epidermidis, Escherichia coli, Bacillus mycoides, and Candida albicans.*

Suspensions of test microorganisms were prepared according to the standard of turbidity (10^9 of cells per 1 ml). They were then introduced into nutrient mediums melted and cooled down to 50°C (0.1 ml of suspension of test organism per 100 ml of medium), which were then distributed in Petri dishes per 20 ml.

In the hardened agar medium, sowed by microorganism under test, holes were made using a sterile drill. We introduced in the holes alcohol or acetone extracts in a concentration 1:10 (0.1 ml). An alcohol/acetone solution in the concentration 1:10— which concentration did not inhibit the growth of test microorganisms—was used as control. Dishes with test microorganisms and extracts under study were kept in thermostat under 25°C during 24 h. If microorganisms are sensitive to antibiotic substances, extracted by alcohol/acetone, then zones without growth appear around holes. We evaluated antimicrobial activity of herb extracts according to the diameter of zone of growth inhibition of the test microorganism [6].

Antibiotic activity relative to test microorganisms was found for alcohol extracts of most of herbs (Table 12.1). All herb extracts under consideration did not show antibiotic activity relatively to *E. coli.*

TABLE 12.1   Antibiotic activity of alcohol extracts of flowers and inflorescences of medicinal herbs relatively to conditionally pathogenic microorganisms

Species of medicinal herb	Diameter of zone of growth inhibition of the test microorganism, $(x \pm m_x)$ (mm)				
	Staphylococcus aureus	Staphylococcus epidermidis	Escherichia coli	Bacillus mycoides	Candida albicans
Potentilla erecta L.	0	13.0 ± 0.00*	0	11.0 ± 0.09*	0
Erungium planum L.	12.1 ± 0.13*	11.9 ± 0.08*	0	11.0 ± 0.02*	0
Prunella vulgaris L.	0	10.0 ± 0.00*	0	0	0
Origanum vulgare L.	0	11.0 ± 0.00*	0	9.9 ± 0.04*	0
Tanacetum vulgare L.	11.0 ± 0.05*	10.1 ± 0.02*	0	11.0 ± 0.00*	0
Agrimonia eupatoria L.	11.0 ± 0.02*	12.1 ± 0.10*	0	10.0 ± 0.00*	0
Nepeta cataria L.	0	7.0 ± 0.00*	0	0	0
Potentilla recta L.	11.2 ± 0.09*	17.1 ± 0.16*	0	9.0 ± 0.00*	$14 \pm 0,11$ *
Leonurus cardiaca L.	0	10.1 ± 0.02*	0	10.0 ± 0.00*	0
Betonica officinalis L.	0	0	0	11.1 ± 0.03*	0
Cichorium intybus L.	0	$11.0 \pm 0,09$*	0	0	0
Symphytum officinale L.	0	$7.1 * 0.02$*	0	0	0
Echinacea purpurea L.	0	6.0 ± 0.00*	0	0	0

Note: *—differences with control are significant with $P < 0.05$; control is the alcohol solution (dissolved 1:10) in which zone of growth inhibition of test organism was absent.

The growth of golden staphylococcus (*S. aureus*) was inhibited by extracts of inflorescences of *Eryngium planum* L., *Tanacetum vulgare* L., *Agrimonia eupatoria* L., *Potentila recta* L., and extracts of leaves of *Tanacetum vulgare* L. and *Glycyrrhiza uralensis* L. Maximum zones of growth inhibition of *St. aureus* were observed for alcohol extracts from leaves of *Tanacetum vulgare* L. and *Glycyrrhiza uralensis L.*: the diameter of zone of growth inhibition of bacteria was for them 13 mm (Table 12.2).

TABLE 12.2　Antibiotic activity of alcohol extracts of leaves of medicinal herbs relatively to conditionally pathogenic microorganisms

Species of medicinal herb	Diameter of zone of growth inhibition of the test microorganism, $(x \pm m_x)$ (mm)				
	Staphylococcus aureus	Staphylococcus epidermidis	Escherichia coli	Bacillus mycoides	Candida albicans
Potentilla erecta L.	0	7.1 ± 0.03*	0	0	0
Filipendula ulmaria L.	0	12.1 ± 0.14*	0	10.9 ± 0.12*	0
Tussilago farfara L.	0	6.1 ± 0.02*	0	0	0
Armoracia rusticana Lam.	0	12.0 ± 0.11*	0	11.9 ± 0.13*	0
Lilium martagon L.	0	0	0	6.0 ± 0.01*	0
Geranium pratense L.	0	20.0 ± 0.16*	0	10.2 ± 0.07*	18.1 ± 0.15*
Erungium planum L.	0	10.0 ± 0.00*	0	11.2 ± 0.16*	0
Calendula officinalis L.	0	10.1 ± 0.02*	0	11.0 ± 0.03*	0
Origanum vulgare L.	0	19.1 ± 0.17*	0	0	0
Fragaria vesca L.	0	0	0	0	14.2 ± 0.11*
Tanacetum vulgare L.	12.9 ± 0.14*	11.0 ± 0.01*	0	11.1 ± 0.03*	0
Agrimonia eupatoria L.	0	12.2 ± 0.11*	0	0	0
Nepeta cataria L.	0	10.9 ± 0.07*	0	9.1 ± 0.04*	0
Potentilla recta L.	0	11.0 ± 0.02*	0	0	0
Leonurus cardiaca L.	0	9.9 ± 0.05*	0	10.0 ± 0.04*	0

TABLE 12.2   *(Continued)*

Species of medicinal herb	Diameter of zone of growth inhibition of the test microorganism, $(x \pm m_x)$ (mm)				
	Staphylococcus aureus	Staphylococcus epidermidis	Escherichia coli	Bacillus mycoides	Candida albicans
Betonica officinalis L.	0	11.2 ± 0.07*	0	10.0 ± 0.02*	0
Cichorium intybus L.	0	0	0	6.1 ± 0.13*	0
Symphytum officinale L.	0	11.2 ± 0.08*	0	0	0
Echinacea purpurea L.	0	0	0	6.0 ± 0.00*	0
Glycyrrhiza uralensis L.	13.0 ± 0.00*	13.2 ± 0.09*	0	15.3 ± 0.11*	0
Urtica cannabina L.	0	6.0 ± 0.00*	0	0	0
Poligonatum officinale All.	0	6.2 ± 0.08*	0	0	0

Note: *—differences with control are significant with P < 0.05; control is the alcohol solution (dissolved 1:10) in which zone of growth inhibition of test organism was absent.

Maximum zones of growth inhibition of *St. epidermidis* were observed for alcohol extracts from leaves of *Geranium pratense* L. and *Origanum vulgare* L., where the diameter of zones of growth inhibition of bacteria was 20 and 19 mm, respectively. Under the effect of alcohol extracts from the flowers of *Potentila recta* L., *Potentila erecta* L., *Eryngium planum* L., *Agrimonia eupatoria* L. and leaves of *Filipendula ulmaria* L., *Armoracia rusticana* Lam., *Agrimonia eupatoria* L., and *Glycyrrhiza uralensis L.* zones of growth inhibition reached 17 mm, 13, 12, 12, 12, 12, 12, and 13 mm, respectively.

Growth of *B. mycoides* was also inhibited by alcohol extracts from leaves and flowers of many herbs under study. Maximum of zone of growth inhibition of *B. mycoides* was observed under the effect of biologically active substances extracted by alcohol from leaves of *Glycyrrhiza uralensis L.* and leaves of *Armoracia rusticana* Lam., for which the diameter of zones of test organism growth inhibition reached 15 and 12 mm, respectively. The growth of *C. albicans* was inhibited by alcohol extracts from leaves of *Geranium pratense* L. and *Fragaria vesca* L. and from the flowers of *Potentila recta* L., when corresponding diameters of zones of test organism growth inhibition reached 18, 14, and 14 mm, respectively.

Experiments showed antibiotic activity of acetone extracts from most of herbs relatively to test organisms. All acetone extracts under study did not show any antibiotic activity relatively to *E. coli*.

The gowth of golden staphylococcus (*S. aureus*) was inhibited by extracts of flowers of *Potentila recta* L., *Origanum vulgare* L., *Fragaria vesca* L., and *Eryngium planum* L and by leaves of *Glycyrrhiza uralensis* L. and *Filipendula ulmaria* L. Maximum zone of growth inhibition *of St. aureus* was found for acetone extract from flowers of *Potentila recta*: the diameter of zone of bacteria growth inhibition reached 13 mm (Table 12.3).

TABLE 12.3   Antibiotic activity of acetone extracts of leaves of medicinal herbs relatively to conditionally pathogenic microorganisms

Medicinal herb species	Diameter of test-microorganism growth inhibition zone, ($x \pm m_x$) (mm)				
	Staphy-lococcus aureus	Staphy-lococcus epider-midis	Esch-erichia coli	Bacillus my-coides	Candida albicans
Inflorescences and flowers					
Potentilla recta L.	13.0 ± 0.00*	16 ± 0.15*	0	14.0 ± 0.11*	13 ± 0.00*
Digitalis grandiflora Mill.	0	0	0	0	0
Hypericum perforatum L.	0	0	0	0	0
Calendula officinalis L.	0	0	0	0	0
Origanum vulgare L.	11.0 ± 0.03*	0	0	0	11.0 ± 0.02*
Leonurus cardiaca L.	0	0	0	0	0
Erungium planum L.	12.0 ± 0.09*	0	0	0	0
Nepeta cataria L.	0	0	0	0	0
Leaves					
Valeriana officinalis L.	0	0	0	0	0

TABLE 12.3   *(Continued)*

Medicinal herb species	Diameter of test-microorganism growth inhibition zone, $(x \pm m_x)$ (mm)				
	Staphylococcus aureus	Staphylococcus epidermidis	Escherichia coli	Bacillus mycoides	Candida albicans
Potentilia recta L.	0	0	0	0	0
Digitalis grandiflora Mill.	0	0	0	0	0
Armoracia rusticana Lam.	0	0	0	0	0
Potentilla erecta L.	0	11.0 ± 0.09*	0	0	12.0 ± 0.10*
Calendula officinalis L.	0	0	0	0	0
Tussilago farfara L.	0	0	0	0	0
Origanum vulgare L.	0	0	0	0	0
Fragaria vesca L.	11.0 ± 0.02*	12.0 ± 0.10*	0	12.0 ± 0.10*	0
Urtica cannabina L.	0	0	0	0	0
Leonurus cardiaca L.	0	0	0	0	0
Filipendula ulmaria L.	12.0 ± 0.09*	14.0 ± 0.11 *	0	11.0 ± 0.05*	11.0 ± 0.00*
Erungium planum L.	0	0	0	0	0
Glycyrrhiza uralensis L.	12.0 ± 0.10*	12.0 ± 0.11*	0	13.0 ± 0.00*	0
Agrimonia eupatoria L.	0	0	0	0	0

Note: *—differences with control are significant with $P < 0.05$; control is the acetone solution (dissolved 1:10) in which zone of growth inhibition of test organism was absent.

Maximum zones of growth inhibition of *St. epidermidis* were found under the effect of biologically active substances extracted by acetone from the flowers of *Potentilia recta* L. and the leaves of *Filipendula ulmaria* L., where the diameter of zones of bacteria growth inhibition reached 16 and 14 mm, respectively.

Inhibition zones for *St. epidermidis* were also observed under the effect of acetone extracts from leaves of *Potentilia erecta* L., *Fragaria vesca* L. and *Glycyrrhiza uralensis* L. –11, 12, and 12 mm, respectively.

The growth of *B. mycoides* was inhibited by acetone extracts from the flowers of *Potentilia recta* L. and from the leaves of *Fragaria vesca* L., *Glycyrrhiza uralensis* L., and *Filipendula ulmaria* L. Maximum zones of growth inhibition of *B. mycoides* were observed under the effect of acetone extracts from flowers of *Potentilia recta* L. and leaves of *Glycyrrhiza uralensis* L., when the diameter of test organism growth inhibition zones reached 14 and 13 mm, respectively.

The growth of *C. albicans* was inhibited by acetone extracts from the flowers of *Potentilia recta* L., leaves of *Potentilia erecta* L., flowers of *Origanum vulgare* L., and leaves of *Filipendula ulmaria* L., when the diameter of test organism growth inhibition zones reached 13, 12, and 11 mm, respectively.

12.3   CONCLUSIONS

1. Our study showed that biologically active substances were presented both in flowers and in leaves of medicinal herbs. The differences in the antibiotic activity of extracts from leaves and flowers of herbs relatively to test microorganisms could be conditioned by the chemical composition of phytoncides, as it is known that in most of cases phytoncides are presented by a complex combination of chemical compounds, characteristic for each herb species. These herbs contain intracellular fractions of phytoncides, which are one of the factors of plant natural immunity.

2. Antibiotic activity of medicinal herb phytoncides relative to cultures of conditionally pathogenic microorganisms, namely *S. aureus, S. epidermidis, B. mycoides, and C. albican depended on the species of herbs under study.* The extracts did not inhibit the growth of *E. coli.*

3. The highest phytoncide activity was shown by alcohol extracts of leaves of *Tanacetum vulgare* L. and *Glycyrrhiza uralensis* L. (relative to *S. aureus*), *Geranium pratense* L. and *Origanum vulgare* L. (relative to *S. epidermidis*), *Geranium pratense* L. and *Fragaria vesca* L., flowers of *Potentilia recta* L. (relative to *C. albican*). Among acetone extracts, the highest antibiotic activity was found for flowers of *Potentilia recta* L. and leaves of *Filipendula ulmaria* L. relative to most of conditionally pathogenic microorganisms (*S. aureus, S. epidermidis, B. mycoides, and C. albican*).

4. Antibiotic activity of alcohol extracts is higher than that of acetone extracts, what can be determined by different chemical composition of phytoncides of medicinal herbs under study.

KEYWORDS

- **Conditionally pathogenic microorganisms**
- **Medical plants**
- **Microbial number**
- **Phytoncyds**

REFERENCES

1. Tokin, B. P.; Phytoncides. Tokin, B. P.; Moscow: Publishing House of Tyumen State University; **1948,** 256 p. (in Russian).

2. Blinkin, S. A.; Phytoncides Around Us. Blinkin, S. A.; and Rudnickaja, T. V.; Moscow: Znanije; **1980,** 144 p. (in Russian).
3. Gorlenko, M. V.; Phytoncides of higher plants and phytopathogenic microorganisms. Gorlenko M. V.; Phytoncides, their Role is in the Wild. Kyiv: Naukova dumka; **1957,** 57–61 p. (in Russian).
4. Kosilapova, L. F.; Microbiology. Kosilapova, L. F.; Methodical Guidance to Laboratory-Practical Employments. Tyumen: Publishing House of Tyumen State University; **2000,** 36 p. (in Russian).
5. Ivanenko, A. S.; Agroclimatic Terms of the Tyumen Area: Train Aid [Text] Ivanenko, A. S.; Kuljasova, O. A.; Tyumen: Tyumen State Agricultural Academy; **2008,** 206 p. (in Russian).
6. Polevoj, V. V.; Phytophysiology. Polevoj, V. V.; Voscow; Higher School; **1989,** 464 p. (in Russian).

CHAPTER 13

REACTION OF WINTER WHEAT ON HYDROTHERMIC FACTORS OF WINTERING AND VEGETATION CONCERNING BIOLOGICAL STABILITY OF PLANTS IN THE CONDITIONS OF SOUTH OF TYUMEN REGION OF RUSSIA

ELENA AL. TYUMENTSEVA, NINA AN. BOME,
and ALEXANDER YA. BOME

CONTENTS

13.1   INTRODUCTION

Wheat is one of the most important food cultures in Russia, as well as across the world. In the agricultural zone of Tyumen region, soft spring wheat is the leading culture among cereals. Area under cereal crop in the region increased from 616 thousand ha in 2007 to 7,155 thousand ha in 2011. At the same time, crop area of winter wheat decreased from 6.6 thousand ha in 2009 to 0.4 thousand ha in 2011. The productivity of winter wheat varied within a wide range—from 9.4 centners per ha in 2007 to 36.7 centners per ha in 2011.

A limited distribution of winter wheat in the regions characterized by complicated soil and climatic conditions has several reasons. Thus, L.V. Vikulova [1], M.V. Nikolayev [2], and L.V. Karpova [3] noticed hard wintering conditions, the possibility of return of frost in spring, and early spring drought events among the main limiting factors.

A.S. Ivanenko, N.A. Ivanenko [4] noticed the necessity to develop the technology of breeding of the given culture. The variety has a considerable importance for the wintering and the increase of productivity [5].

It is necessary to note that the winter wheat has a set of advantages compared with spring wheat: it withstands better the early summer drought, it uses more effectively the moisture from summer and autumn precipitation, it suppress the growth of weeds, and the foundation and formation of yield occur under more favorable conditions [6, 7].

It is possible to distinguish several critical periods relatively to environmental factors in the ontogenesis of a plant of winter wheat. Usually, the plants are most sensitive to unfavorable environmental factors at the stage of seedlings and at the first stages of growth [8]. At this stage, the following limiting factors can be noted: sudden changes of air temperature and amount of precipitation (initial stages of growth and development of plants can occur both, under overwatering as well as under insufficient water supply).

During autumn period, the formation of plant biomass and accumulation of nutrients for further wintering takes place. For successful passage of the phase of third leaf, the plants need optimal temperature regime: not above +7–12°C (especially under insufficient moisture) and not below +2–3°C. In winter, the soil temperature at the depth of the node of bushing out affects considerably the vitality of plants. Under the temperature below −19°C, the die-back of point of increase is possible, which leads to the death of the whole plant. The temperature decrease is especially dangerous for the plants of winter wheat from mid-November to early December under small depth of snow cover. The temperature oscillations (interchange of thaws and frosts) can lead to the deformation and breach of root system of plants.

The death of plants in spring (period of vegetation resumption) may be evoked by the return of cold weather; overwatering and hypoxia in the depressions of relief; damage of plants by phytopathogenic fungi; and water deficit.

Each ecological niche has its own peculiarities, which may considerably affect the growth and development of plants. When testing the samples of winter wheat in the northern forest-steppe of Tyumen region from 1987 to 1992, we observed that the death of plants often falls to the spring period due to the complexity of unfavorable factors, among which low temperature is the most important factor. In the first half of May, soil frost down to −5°C and sometimes even lower is possible. Usually, it affects more the varieties of winter wheat, which renew the spring vegetation earlier [9].

In concrete conditions varieties with high adaptive properties, created or chosen for a particular zone, can show their advantages, which is proved by practice, both in Russia, as well as abroad, already a longtime ago.

Access to the tube and earing, when intensive growth of vegetative organs and the formation of generative organs take place, is a critical period from the viewpoint of water supply for the plants of winter wheat [10].

Hence, the selection of varieties and creation of initial material with high eco-logical plasticity remains relevant for the south of Tyumen region. To determine the adaptive potential of the culture and revelation of characters valuable for selection, a complex evaluation of winter wheat samples from the world collection of N.I. Vavilov Research Institute of Plant Industry is carried out.

The aim of the present study is the research of biological stability of winter wheat plants relatively to environmental factors.

The particular tasks include the following:

– Analyzing the biological stability of plants of winter wheat during five vegeta-tion seasons (2007–2011);
– Comparing the reaction of samples to meteorological factors according to the character of plant survival; selecting the samples with stable display of this character.

13.2 MATERIALS AND METHODOLOGY

A total of 118 collection samples of winter wheat, differing by origin, were tested; the most of samples came from Krasnodar krai (13.6%), Samara region (7.6%), the United States (10.9%), Ukraine (11.8%), and samples from foreign selection amounted 35.6 percent.

Based on morphological characters (color of ear and caryopsis, presence or ab-sence of aristas, pubescence of ear scales), samples were attributed to six botanical varieties: *lutescens* (Alef.) Mansf., *albidum* Al., *graecum* (Koern.) Mansf., *erythro-spermum* Korn., *ferrugineum* (Alef.) Mansf., *velutinum* Schubl.

Field tests of collection samples were carried out in 2007–2011 at the experimental proof biological station "Lake Kuchak" of Tyumen State University (Nizhnetavdinsk district of Tyumen region). The territory is moderately wetted, hydrothermal coef-ficient is 1.2–1.3. Annual precipitation total is 365 mm. Sum of positive temperatures above 10°C is 1875 degree days [11]. Soil is cultivated sod-podzolic.

Sowing of samples was performed manually in late August, The plots of land con-sisted of three rows. Each row was seeded at 200 seeds. The length of row was 1 m, width of row spacing was 15 cm, and seeds were placed at the depth 5–6 cm. Allot-ments were placed in tiers. Plants were harvested manually by pulling out with the root.

Phenological observations and morphological description were performed accord-ing to methodic guidelines for the research of world wheat collection, [12] Interna-tional CMEA classifier of genus *Triticum* L. [13], and methodic guidelines (replenishment, keeping alive, and research of world collection of wheat, aegilops and triticale) [14].

Biological stability characterizes the reaction of plant organism during its onto-genesis to combined effect of environmental factors [15]. The amount of plants, which passed all phonological phases until ripeness, is the criteria of stability.

13.3   RESULTS AND DISCUSSION

The years of study differed by moisture supply and temperature regime both among them, as well as from long-term means. Thus, vegetation seasons 2007 and 2008 were wet and warm; in 2009, 2010, and 2011 water deficit was observed at the background of elevated air temperature (in 2010 and 2011).

The temperature of December–February 2007–2008 exceeded long-term mean by 3.0–3.5°C. Precipitation total in 2007 exceeded normal (115%); and in 2008, it consisted 68.5 percent of normal. Winters 2009, 2010, and 2011 were cold; but in 2009–2010, it was a lot of snow and in winter 2011 it was only little snow.

In these conditions, the survival probability of plants of the studied winter wheat samples varied considerably (from 10.5 to 100.0%); overall mean for all years was 53.0 percent (Table 13.1). High values of the coefficient of variation (32.09–41.55%) indicate high variability of the character.

TABLE 13.1   Variability of survival probability of plants of the winter wheat samples during years differing by meteorological conditions

Year of study	Mean of samples (%)	CV (%)	Min	Max
2007	48.4 ± 2.39•	41.55	13.0	100.0
2008	75.5 ± 2.29*•	32.09	13.1	100.0
2009	55.1 ± 2.31•	39.48	11.6	100.0
2010	49.7 ± 1.76•	37.76	16.2	100.0
2011	35.8 ± 1.34*	38.02	10.5	69.4
5-years mean	53.0 ± 1.09	37.78	12.5	82.6

Note: statistically significant differences (P < 0.05) with ·—2011; *—5-years mean for all samples; and CV, %—coefficient of variation.

The lowest probability of plant survival was observed 2011, which can be related with insufficient amount of precipitation in December–February (41.5% of normal). The maximum probability of survival (75.5%) was observed in the wettest year from the whole period—in 2008, when annual precipitation total was 539.3 mm compared with long-term average of 382.9 mm (141.0% of normal). Plant water supply could be considered as sufficient, as a high precipitation amount was in February–March, exceeding long-term average by 76.0 percent; and in May during growth of plants, it was optimal (102.0% of normal).

When dividing samples conditionally into four groups according to their probability of survival it was found that on average during the years of the study the majority of samples varied from 31.0 to 60.0 percent. This regularity was observed for 4 years from 5, excluding 2008, when samples with high and very high values of the character prevailed (Figure 13.1).

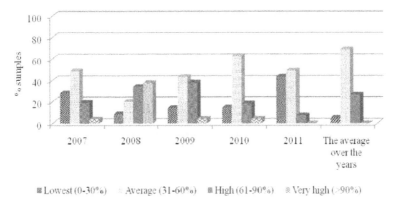

FIGURE 13.1   The distribution of winter wheat survival.

As a result of 5-years' testing samples with the minimum coefficient of variation relatively to others, from 7.0 to 27.0 percent, i.e., with less-expressed reaction to the changing environmental conditions, were selected from the collection of winter wheat. In average for the whole period under study 12 samples showed interannually stable biological stability (above 50%), in four samples this characteristic was stably low. According to the origin, the samples with relatively stable display of the character were from the regions of Russian Federation (Saratov, Orenburg, Rostov, Omsk and Novosibirsk regions, Krasnodar krai, Tatarstan), as well as from foreign countries (the United States and Ukraine).

Generalization and comparison of experimental data according to the combined effect of environmental factors showed that the probability of survival is higher in the samples from different soil and climatic zones of Russia (Table 13.2).

TABLE 13.2   Comparative characteristic of domestic and foreign samples according to the probability of survival

Year of study	Domestic				
	Min	Max	$x \pm S_x$ (cm)	CV (%)	Span
2007	17,3	100,0	$54,7 \pm 3,00$	32,87	82,7
2008	30,4	100,0	$77,3 \pm 2,73*$	28,70	69,6
2009	11,6	100,0	$57,1 \pm 3,16$	40,69	88,4

TABLE 13.2 *(Continued)*

Year of study	Min	Max	Domestic $x \pm S_x$ (cm)	CV (%)	Span
2010	20,6	100,0	53,9 ± 2,41	36,62	79,4
2011	18,2	66,3	35,9 ± 1,38*	30,43	48,1
5-years mean	27,7	82,6	55,8 ± 1,27	33,86	54,9
Foreign					
2007	13,0	100,0	42,0 ± 3,45	48,61	87,0
2008	16,0	100,0	76,7 ± 3,84*	31,28	84,0
2009	25,0	79,4	48,2 ± 3,26	36,97	54,4
2010	16,2	88,7	44,3 ± 2,53	36,61	72,5
2011	10,5	69,4	34,7 ± 2,96*	49,77	58,9
5-years mean	22,3	66,1	49,1 ± 1,81	40,52	

Note: *—statistically significant difference with 5-years average (P < 0.05)

The variability of the character for averaged data of the collection was high, both for domestic and foreign samples. Generally, in the whole collection, minimum probability of survival was observed in 2011 and the maximum in 2008 independently on the origin of samples. When comparing domestic and foreign samples according to the span of the character in different years, no regularity was found.

The data obtained can be applied as a theoretic base of the formation of initial selection material of winter wheat and creation of varieties with broad ecological plasticity.

It is known that the cultivation technology considerably affects the realization of potential properties of a variety. Our research showed that the effect of unfavorable factors on the plants of winter wheat may be reduced under sowing on the link fallow. Link culture is mustard, and its optimal term of sowing is July 12–15. Links are placed every 4.5 m. When sowing winter wheat in August 15–25, the height of links reached 32–36 cm, and before wintering it reached 97–132 cm. It is necessary to note that in spring underreturning of cold links may carry protection function for the plants of winter wheat, providing a special microclimate because of slow snow melt and wind protection. The application of links allowed the increase of the stocking density of plants per unit area after wintering; spring growth of plants was more intensive [9].

Despite the fact that there is a contradiction between stability and productivity, it could be overcome by the following components: by means of the research and conservation of hereditary diversity of gene pool of winter wheat and permanent improvement of adaptive technologies of cultivation including the development of nontraditional approaches, compensating negative properties of varieties that we did not succeed to eliminate during selection.

13.3 CONCLUSIONS

It was shown in the research that the heterogenic material of winter wheat, which included samples of different ecological and geographic origin, provides the necessary ecological plasticity of the culture.

Generally, adaptive potential of winter wheat can be evaluated according to the norm of reaction of plants on the changing environmental conditions during several years with different meteorological factors.

Collection samples of winter wheat under study differed according to biological stability of plants during wintering and vegetation in response to meteorological factors.

Maximum and minimum amounts of plants, which reached full ripeness, were observed in 2007–2008 and in 2010–2011, respectively.

The character of probability of plant survival during vegetation season showed high variability in all years of the study. For most of samples, the probability of survival varied from 31.0 to 60.0 percent, excluding year 2008 with higher values of the character (61–100%).

KEYWORDS

- **Character**
- **Environment**
- **Probability of survival**
- **Variability**
- **Winter wheat**

REFERENCES

1. Vikulova, L. V.; Winter Crops in the North Ural. Vikulova, L. V.; Novosibirsk: Siberian Branch of the Russian Academy of Agricultural, Agricultural Research Institute of Northern Zauralye. **2006,** 232 p. (in Russian).
2. Nikolajev, M. V.; Modern Climate and Changeability of Harvests. Nikolajev, M. V.; Saint Petersburg: Gidrometeoizdat; **1994,** 199 p. (in Russian).
3. Karpova, L. V.; Productivity of Winter Wheat at the Different Terms of Sowing. Karpova, L. V.; Grain Growing; **2005,** *4,* 26–29 p. (in Russian).
4. Ivanenko, A. S.; A Winter Wheat and Triticale are Powerful Reserve of Increase of the Productivity of the Fields of the Tyumen Area. Ivanenko, A. S.; and Ivanenko, N. A.; Agrarian Announcer of Ural; **2012,** *9(101),* 6–7 p. (in Russian).
5. Ismagilov, R. R.; Features of Till of Winter Wheat in the Conditions of Republic of Bashkortostan. Ismailov, R. R.; Gaifullin, R. R.; Nurlygajanov, R. B.; Grain Growing; **2005,** *4,* 29–31 p. (in Russian).
6. Shevelukha, V. S.; State and prospects of selection, seed-grower and intensive technology of till of winter wheat. Shevelukha, V. S.; Vasilenko, I. I.; and Semenova, T. N.; Selection, Seed-Grower and Intensive Technology of Till of Winter Wheat. Collection of Scientific Works. Moscow: Agropromizdat; **1989,** 3–11. (in Russian).

7. Plant-Grower with Bases of Selection and Seed-Grower. Korenev, Rd. G. V.; Moscow: Agro-promizdat; **1990,** 575 p. (in Russian).

8. Zhuchenko, A. A.; Adaptive Potential of Cultural Plants (Environmental and Genetic Bases). Zhuchenko, A. A.; Kishinev: Shtiinca; **1988,** 767 p. (in Russian).

9. Bome, N. A.; Phenotypical realization of signs of spring and winter-annual wheat is in North Urals. Bome, N. A.; Bome, Ya. A.; Announcer of the Tyumen State University. Tyumen: Publishing House of the Tyumen State University; 2005, 5, 256–263 (in Russian).

10. Zykin, V. A.; Ecology of Wheat: Monograph. Zykin, V. A.; Shamanin, V. P.; and Belan, I. A.; Omsk: Publishing House of Omsk state Agrarian University; **2000,** 124 p. (in Russian).

11. Ivanenko, A. S.; Agroclimatic terms of the Tyumen area. Ivanenko, A. S.; Kuljasova, O. A.; Tyumen: Publishing House of Tyumen State Agricultural Academy; **2008,** 206 p. (in Russian).

12. Gradchaninova, O. D.; Methodical Pointing on the Study of World Collection of Wheat. Gradchaninova, O. D.; Filatenko, A. A.; and Rudemko, M. I.; Leningrad: VIR; **1987,** 28 p. (in Russian).

13. International Classifier Comecom of Family *Triticum* L. Leningrad; **1984,** 84 p. (in Russian).

14. Merezhko, A. F.; Addition, maintenance in a living kind and study of world collection of wheat, aegilops, and triticale. Merezhko, A. F.; Udachim, R. A.; Zujev, V. E.; et al. Methodical Pointing. St Petersburg: VIR; **1999,** 82 p. (in Russian).

15. Guzhov, Yu. L.; Selection and Seed-Grower of Cultural Plants. Guzhov, Yu. L.; Fux, A.; and Valichek, P.; Moscow: Mir; **2003,** 539 p. (in Russian).

PART III
PLANT BREEDING IN THE UKRAINE: AGROECOLOGICAL APPROACHES

CHAPTER 14

THE STUDY OF THE ANATOMICAL STRUCTURE OF APPLE-TREE TISSUES AND APPLE FRUITS (*MALLUS* MILL.)

VOLODYMYR V. ZAMORSKYI and ANATOLY IV. OPALKO

CONTENTS

14.1   INTRODUCTION

About 75 percent of apple trees are grown in commercial orchards of Ukraine, and 40 percent are grown in home orchards. Past decades witness a positive tendency in apple fruits production in compliance with a national program of horticulture development in Ukraine approved in 2008; it is planned to increase apple orchards to 144.8 thousand ha for the period up to the year of 2025 [1].

The genus *Malus* was first summarized in pre-Linnaean literature by Tournefort (1700) [2, 3]. In «Species plantarum...», Linneaus (1753) described this genus as species of the genus *Pyrus* [4] and was later adopted by Miller (1754) as genus separate [5]. Since then genus *Malus* Mill. Was classified as a member of the family Rosaceae, subfamily Maloideae [6].

Results of investigations done by the scientists of different countries, including molecular genetic studies as to genus *Malus* and their closest congeners, served as a basis for a revised classification system for genus and its taxonomic position within the Rosaceae Juss. Family [3]. Actually, the apple tree belongs to the genus *Malus* subtribe Pyrinae, tribe Pyreae, subfamily Spiraeoideae (formerly Maloideae), family Rosaceae [7]. The present *Malus* taxonomy includes 35 accepted species names, and 13 names of infraspecific rank [8, 9].

Most of the wild apple species were found in the mountains of central and inner Asia, western and southwestern China, Far East, and Siberia. Just at that spot, southwestern China and inner and central Asia are situated as the largest center of diversity of the genus *Malus* in general [2, 10].

A wide use of grafting as the main method of growing planting materials of an apple-tree cultivars causes the necessity to study the structural processes in calluses, which define the joining during the process of grafting. Until now, only general principles of the processes due to insufficient studying of anatomic aspect of grafting methods are given in current horticultural literature. While studying grafted young apple trees, special attention is paid to the place of joining rootstock and scion, which plays a crucial role in the general movement of substances, such as ions, water, and growth-regulating hormones, in xylem [11–13].

Fruit of an apple develops from a lower fruit-set; its pulp is formed of beyond-carpel part of the fruit. There is an opinion, Esau [11], that beyond-carpel tissue has appendicular origin (flower tube or hypanthium). Some botanists consider additional tissue to be part of receptacle; they call the external juicy part of the fruit a skin. To identify fruit structure and flavor properties of fresh cut produce, we studied apple fruits grown in the Forest Steppe Zone of Ukraine [14].

14.2   MATERIALS AND METHODOLOGY

Young trees of *Malus* domestica Borkh., of different age and grafting method, grown in the nursery of the department of horticulture at Uman National University of Horticulture, were chosen for studying. In some variants, leading elements of the xylem were painted with a dye using our own method.

Anatomical structure was studied according to the methodology worked out by A.O. Hrytsaienko [14–16] and our observations made with the help of updated methods of computer-aided fixing of the cuts made by video-attachment 'Philips ToUcam camera' and measurement of the sizes of fruit structural elements with the help of special computer-aided system 'Image Scope Lite' assigned for microscopy and analysis.

Apple fruits of such cultivars as 'Golden Delicious,' 'Aidared,' 'Jonathan,' 'Mantuaner,' 'Melrose,' 'Rubinove Duky' were taken from an experimental orchard of Uman National University of Horticulture.

Nonirrigated orchard was planted on rootstock M.9 and on seedling rootstock 'Antonivka' with the use of inter-stem PB 62-396. Anatomical structure was studied according to the methodology worked out by A.O. Hrytsaienko [15] and our observations made with the help of updated methods of computer-aided fixing of the cuts made by video-attachment 'Philips ToUcam camera' and measurement of the sizes of fruit structural elements with the help of special computer-aided system 'Image Scope Lite' assigned for microscopy and analysis.

14.3   RESULTS AND DISCUSSION

A lower part of a longitudinal cut of grafting with one part of callus tissue, which was formed between secondary xylem of rootstock M26 and hydrocide bud of cultivar 'Johnagold,' is shown (Figure 14.1).

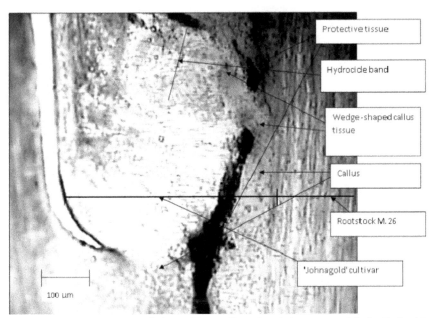

FIGURE 14.1   Longitudinal radial cut of the grafting place (a lower part of the bud) of cultivar 'Johnagold' on rootstock M.26.

Such wedge-shaped tissue in a longitudinal section covers (in the form of a ring) the whole section of the tree. A rather wide layer of a protective tissue is seen between a bud and a section of the rootstock; it is formed where grafting components join. It is worth mentioning that a protective tissue, which consists of dead cells and is formed in the place of wound near callus, remains without structural changes during the existence of a grafted plant and is not the spot where rootstock and a bud copulate.

The structure of a protective tissue is as shown in Figure 14.2. It consists of over ten layers of elongated (up to 100 μm) flat cells, which in this case are the remaining elements of a leading rootstock system. Callus in the spot of copulation of a cultivar bud and rootstock has typical yellow color, and it consists of parenchyma cells of various size.

Analyzing Figures 14.1 and 14.2, when grafting is done between a bud shield and rootstock section, leading system elements are formed only at a joint of callus formations. In fact, copulation of separate callus tissues occurs. Simultaneously, the process of cell differentiation starts on some spots of callus tissue.

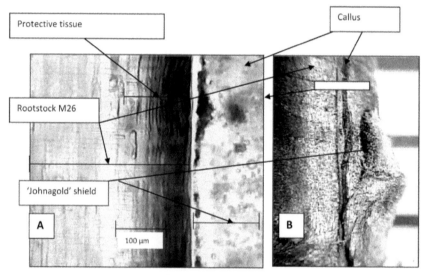

FIGURE 14.2 Longitudinal tangential cut (A) of the middle of the grafting place of cultivar 'Johnagold' on rootstock M.26 increased by 140 times; (B)—a general view of grafting.

The highest viability was typical for the cells, which were formed of cambium and pericycle. Cross- and longitudinal sections of the grafting spot of vigorous apple cultivar 'Gloster' on dwarf rootstock M9 after spring occurrence of a bud during intensive growth and coloring of a leading system in methylene blue are shown in Figure 14.3. There (Figures 14.3(A) and 14.3(B)) one can see arteries of water with dissolved nutrition elements in it, coming to a grafted bud. In a rootstock, elements of secondary xylem and leading elements of the wood perform this function, whereas in a grafted bud newly formed wedge-shaped leading elements do this. Where intensive callus is formed (Figure 14.3(C)), cells of callus tissue have systemless arrangement. However, in the zone of newly formed leading elements and while moving to them, they are located in regular elongated rows.

As a result, the first elements of a leading system—tracheides—are formed from meristem spots in separate areas of callus, and near them closer to primary bark—sieve-shaped tubes. Similar tracheid-shaped cells belong to a widely spread type of water-carrying elements, called hydroxides. Callus hydroxides differ from water-carrying anatomical elements of other types: cell nuclei are kept longer in them; cell thickening lignifies though [17].

As Figure 14.3 shows, the process of changing callus cells into hydrocides, which starts deep in the wedge, spreads and gradually reaches the areas that are closer to callus periphery, forming ramification in all directions. Finally, not far from the periphery each band, which consists of hydorcides and accompanying sieve-shaped tubes, has accumulated hydrocides at the end; they are alternated with the cells that have thick cytoplasmatic content that resembles meristem. The accumulation of hydrocides and meristem cells looks like wedge-shaped nodes (Figure 14.3(C)).

In Figure 14.3, one can see that a protective tissue in the place of grafting was turned back toward the bud. This proves that out of the two callus layers where differentiation processes take place, the one that belongs to rootstock is stronger and cell differentiation occurs more energetically there. The process under consideration, which identifies the contact of vascular systems, is the most crucial for successful grafting. However, the width of hydrocide band, which performs water supply, is rather small and ranges from 0.3 mm in the place, where callus cells of the rootstock and grafted bud join, to 3 mm in the area of intensive bud growth.

This narrow hydrocide band is located near a grafting periphery; that is why extended constriction of the area with polymer materials used for grafting can have a negative effect on root stock water supply.

Among the well-known methods of vegetative propagation of apple trees, grafting (or bud grafting) and improved copulation are of great practical importance. The latter is used for fast growing of young apple trees [18]. Studying the grafting place of one-year-old apple trees, using the above-mentioned methods, on longitudinal radial cut, showed (Figure 14.4(A)) that the use of improved copulation method enabled the joining process of xylem and phloem elements.

Looking at a cross-section projection of the grafting place, one can see a continuous ring-shaped system of leading elements, which ensures both a reliable joint of scion and rootstock, and proper movement of nutrition elements and water. A protective tissue, formed in the joint places of grafting elements, is quite small; it is located in the central part of the young tree and belongs to core elements.

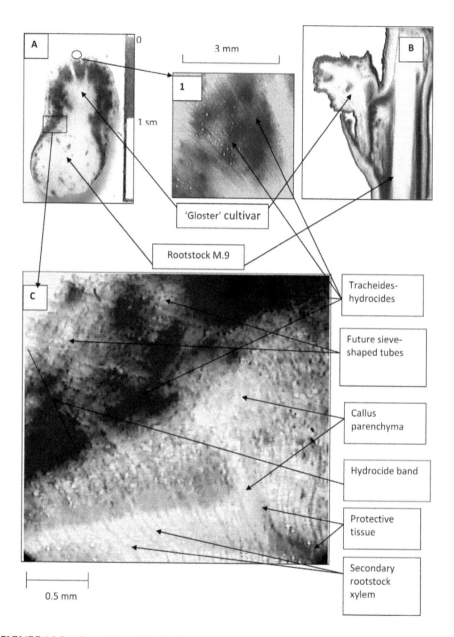

FIGURE 14.3 Cross (A) and longitudinal radial (B) cuts of the grafting place of 'Gloster'/M.9 after bud growing in the period of intensive growth when a leading system was colored in methylene blue. The increase of cross-section: A, B—by 10 and C—by 38 times.

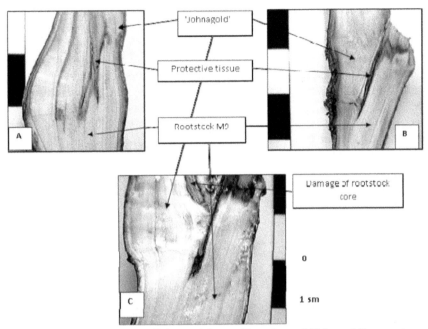

FIGURE 14.4 Longitudinal radial cut of the grafting place of 'Johnagold' young tress on rootstock M.9 depending on the grafting method: A—a 1-year-old grafting using the improved copulation method; B—a 1-year-old grafting; C—a 2-year-old grafting.

Using a grafting method (Figure 14.4(B)), the joining of grafting components occurs in a lower part of the bud (see Figure 14.1), and a leading system is formed only on one side when we look at a longitudinal radial cut of the grafting place. A protective tissue covers a greater part of the grafting place, which is situated between its components and in the upper part of the rootstock; it does not facilitate their reliable growth. Water supply goes through a narrow (up to 0.5 cm) canal of leading elements of secondary xylem and phloem. When time progresses, the area of active movement increases (Figure 14.4(C)), and a protective tissue zone shifts to the upper part of the young tree, being less productive as to the strength and exchange of plastic substances. Attention should be paid to the core damage of a grafted cultivar in a 2-year-old young tree (Figure 14.4(C)), which is not typical to a 1-year-old young tree. Probably, it is due to a weak supply of the rootstock with plastic substances and the effect of unfavorable environmental conditions.

Anatomical studies of the grafting place of young apple trees of various age show essential tissue disorganizations, especially in the xylem area, proved by several trials [19]. Similar changes affect the growth and development of a fruit tree. Some scientists [20] believe that the tissue in the grafting place has an effect on vegetative growth of a shoot, limiting water flow from the root to the shoot, or by moving the substance, particularly useful mineral substances and growth-regulating hormones (most likely cytokinine) from transporting flow.

Such anatomical changes may occur because of the limitation of polar auxin transportation (IAA) through the grafting place and its accumulation in grafting [21]. Polar auxin is a key regulator of xylem differentiation and division in cambial area and an initiator of secondary vascular differentiation. The suggested results show (Figures 14.5(A) and 14.5(C)) that to overcome anatomical dysfunction, connected with xylem structure of transplant tissue, a grafting place has an increase. C.J. Atkinson with other authors [22, 23] state that this is a typical auxin-answer, which is most likely formed due to instability caused by large transport of base petal auxin in the rootstock compared with the transport in a grafted rootstock. As a result, auxin is accumulated in transplant tissues, which explains increased callusogenesis.

Epidermis of apple fruit (Figure 14.5) consists of several cell layers, where the tangential diameter exceeds the radial diameter, lower epidermis cells are of larger size. New layers are clearly seen on epidermis surface, and they have complicated structure. A fruit surface is covered with cutin (Figure 14.5(A)).

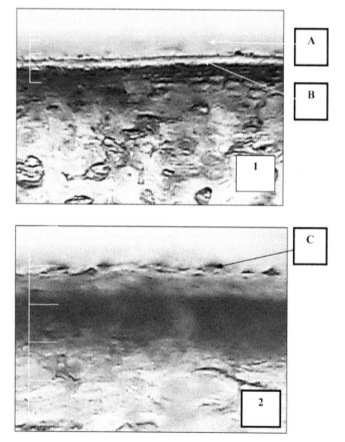

FIGURE 14.5   Structure of apple fruit epidermis (cv. 'Rubomove Duky') on a radial cut. 1—growing with inter-stem M9; 2—growing without inter-stem (control). A—cutin; B—soft wax; C—grains of firm wax. The value of one mark is 100 μк.

According to our researches and other data [24, 25], it consists of such lipid acids as trioxystearine, oxystearine, and dioxypalmitin. A layer of soft wax is located under cutin (Figure 14.5(B)), when a detailed increase is seen (Figure 14.5(C)); firm wax grains are fixed on an uneven surface.

The analysis of the cuts and surface of the apple fruits shows that dense sub-epidermal tissue is developed in external layers; its cells have thick walls and form "a skin." Such cultivars as 'Mantuaner' (350 μк), 'Rubinove Duky' (220 μк), 'Johnathan' (180 μк) have especially thick layer. 'Golden Delicious' fruits have the thinnest "skin" (about 110 μк); in fact, cuticular layer is not observed; soft wax layer is uneven with a great number of gaps. Such structure of superficial layers of apple fruits (ctv. 'Golden Delicious') explains its characteristic ability to lose moisture during long harvesting, and as a result, its marketable properties decrease. It has been recorded [24] that cyanide is located in a colored part of apple fruit epidermis, which is one of the most widely known anthocyanins.

Anthocyanins have violet coloring; however, with ions of metals K, Na, Fe they create compounds of blue color; red coloring is due to the formation of complex compounds anthocyanins with acids (phosphoric and others) [14, 24]. The analysis of cell layers are cross-cut, and superficial part of colored and achromatic apple fruits shows that coloring ranges from light green, green-yellow, pink, to light-red, red, dark red, light-violet. Such spectrum mostly likely results from the condition of complex anthocyanin compounds and adaptive changes of membrane cells due to the increase of protoplast cell surface when fruits accumulate and deposit various substances.

Thorough studying of epidermis cell structure proves that by changing the coloring from light-green to red structural, changes occur in chloroplasts, which have a specific lamellar system. The comparison of colored and achromatic cells (Figure 14.6) shows that when anthocyanins accumulate, changes of internal cell structure occur.

The data received when cell structure of apple fruit skin were increased maximally with help of light microscope (Figure 14.6), and photographs taken by an electronic microscope [14] show that when fruits mature, tilakoides gran and tilakoide strome get wider due to the accumulation of osmiophilic material. In addition, the majority of components with various electronic density, form, and size have triangle and trapezoidal shape. Sometimes, tilakoides widen at the ends and obtain the shape of a barbell. Certain thickening is observed in cell walls of 'Idared' apple skin (up to 6 μк).

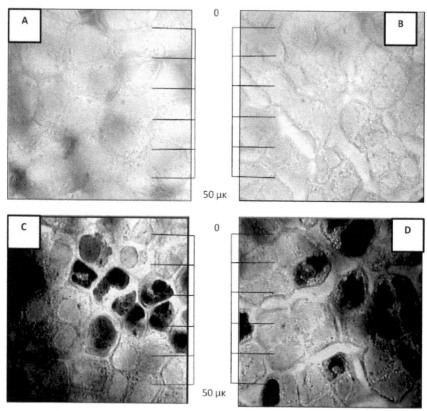

FIGURE 14.6   Structure of achromatic (A, B) and colored (C, D) apple fruit skin cultivar 'Idared'. A, C—without inter-stem (seedling rootstock 'Antonivka'). B, D—inter-stem PB 62—396. The value of one mark is 10 μκ.

The basis of apple fruit is parenchyma saturated with inter-cell spaces, which is formed of huge cells, filled with starch (Figure 14.7).

The analysis of the parenchyma structure of apple fruits (culture of the Forest Steppe Zone of Ukraine) reveals that its structure depends on a cultivar. Thus, parenchyma of such cultivars as 'Idared,' 'Mantuaner,' 'Johnathan' consists of cells whose size ranges from 10 to 110 μκ, inter-cell spaces are dense, no gaps among cells (at a harvesting stage); thus, such fruits have pungent flavor. Parenchyma of cultivars 'Golden Delicious,' 'Melrose' have both small (5–10 μκ) and large (over 100 μκ) cells; they also have tiny cells (up to 5 μκ), which are situated in inter-cell spaces. Due to such structure and the availability of inter-cell spaces, filled with air, fruit parenchyma has soft flavor. Tasting estimation of fresh cut fruits of the above-mentioned cultivar had the best grades.

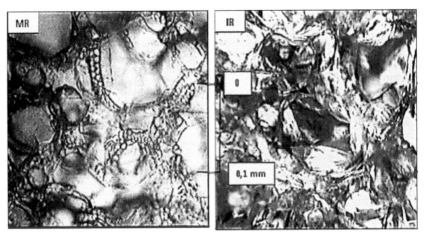

FIGURE 14.7   Structure of apple fruit mesocarpy of various cultivars: MR—'Melrose'; IR—'Idared'. The value of one mark is 100 μк.

Studying apple fruit epidermis of different diameter cv. 'Rubinove Duky' demonstrates that its thickness, when apple fruit diameter is 55 … 78 mm, does not differ as to size and structure. Differences are observed only in sizes of parenchyma cells of pulp: fruits of bigger diameter have larger sizes.

14.4   CONCLUSIONS

Using the method of improved copulation between rootstock and scion facilitates the formation of the system of leading elements, which ensures a reliable joint of rootstock and scion and sufficient movement of nutrition elements and water. In this technology of grafting, water supply occurs via a narrow canal of leading elements in the lower part of grafting, whereas in the upper part nonproductive area of a protective tissue is formed. Probably the improvement of the grafting technology (a detailed control over weakening and removing of isolating material) will result in eliminating the identified drawbacks and getting planting material of better quality; finally, it will lead to the increase of productive potential of apple tree plantations.

Thus, anatomical researches of apple fruits proved that epidermis structure and that of parenchyma cells could be explained by "variety" peculiarities. Epidermis thickness did not depend on the size of apple fruit. Flavor properties of fresh cut apple fruits depended on anatomical structure of parenchyma cells.

ACKNOWLEDGMENT

We thank Tetiana Suhomeilo, a senior teacher of Uman National University of Horticulture, for her technical assistance in translating this chapter.

KEYWORDS

- **Apple skin**
- **Cell**
- **Commercial orchards**
- **Cultivar**
- **Cuticular layer**
- **Horticulture**
- **Lipid acids**
- **Pungent flavor**

REFERENCES

1. Luzan Yu. J.; et al.; Horticulture Development Sectoral Programme of Ukraine for the Period Until 2025. Kyiv: Zhytelev; **2008,** *76* c. (in Ukrainian).
2. Opalko, A. I.; Kucher, N. M.; Opalko, O. A.; and Chernenko, A. D.; Phylogeny and phytogeography pome fruits. Autochthonous and alien plants. The collection of proceedings of the National dendrological park "Sofiyivka" of NAS of Ukraine. **2012,** *8,* 35–44. (in Ukrainian).
3. Opalko, O. A.; Chernenko, A. D.; and Opalko, A. I.; Phylogenetic relationships in the representatives of genus *Malus* Mill. cultivated in Ukraine. *Plants Introduction.* **2012,** *1,* 16–23. (in Ukrainian).
4. Linnaei, C.; *Pyrus.* Species plantarum, exhibentes plantas rite cognitas, ad genera relatas, cum differentiis specificis, nominibus trivialibus, synonymis selectis, locis natalibus, secundum systema sexuale digestas. *Holmiae: Laurentii Salvii.* **1753,** *1,* 479–480.
5. Miller, Ph.; The Gardeners Dictionary [Abridged 4 edition] London: Rivington; **1754,** 558 p.
6. Potter, D.; Gao, F.; Bortiri, P. E.; Oh, S.-H.; Baggett, S.; Phylogenetic relationships in Rosaceae inferred from chloroplast matK and trnLtrnF nucleotide sequence data. *Plant Syst. Evol.* **2002,** *231,* 77–89.
7. Potter, D.; et al., *Phylogeny and classification of rosaceae. Plant Syst. Evol.* **2007,** *266,* 5–43.
8. The Plant List by the Royal Botanic Gardens Kew and Missouri Botanical. **2010,** URL: http://www.theplantlist.org/tpl/search?q=Malus
9. Ignatov, A.; and Bodishevskaya, A.; *Malus.* Wild crop relatives: genomic and breeding resources, temperate fruits. Ed. Chittaranjan Kole. Berlin: Heidelberg: Springer; **2011,** *3,* 45–64.
10. Juniper, B. E.; Mabberley, D. J.; The Story of the Apple. Portland: Timber press; **2006,** 219 p.
11. Esau, K.; Anatomy of Seed Plant. 2nd edition. New York: John Wiley and Sons; **1977,** 576 p.
12. Evert, R. F.; Esau's Plant Anatomy: Meristems, Cells, and Tissues of the Plant Body: Their Structure, Function, and Development. 3rd edition. Hoboken, New Jersey: John Wiley and Sons; **2006,** 607 p.
13. Rudall, P. J.; Anatomy of Flowering Plants An Introduction to Structure and Development. New York: Cambridge University Press; **2007,** 159 p.
14. Zamorskyi, V.; The role of the anatomical structure of apple fruits as fresh cut produce. *Acta Hort. (ISHS).* **2007,** *746,* 509–512.
15. Hrytsaienko, A. O.; Physiological-Biochemical Studying of Apple-Trees with Various Soil Management Systems Used in the Orchard: Dissertation of Candidate of Science (Biol). Uman; **1968,** 88–89 p. (in Russian).
16. Hrytsaienko, Z. M.; Hrytsaienko, A. O.; and Karpenko, V. P.; Methods of Biological and Agro-Chemical Studying of Crops and Soils. Kyiv: NICHLAVA Ltd.; **2003,** 130–132. (in Ukrainian).
17. Aleksandrov, V. H.; Plant Anatomy. M.: Vyschaia Shkola; **1966,** 264.

18. Zamorskyi, V. V.; Method of fast growing of young trees. Declarative patent. A.c. 44495 A. Announced on 24.04. 2001 Decision of "Ukrpatent" of November 29, 2001. Bul. No 2, 15.02.2002. (in Ukrainian).
19. Soumelidou, K.; Battey, N. H.; John, P.; and Barnett, J. R.; The anatomy of the developing bud union and its relationship to dwarfing in apple. *Ann. Bot.* **1994,** *74,* 605–611.
20. Jones, O. P.; Endogenous growth regulators and rootstock/scion interactions in apple and cherry trees. *Acta Hort. (ISHS).* **1986,** *179,* 177–183.
21. Soumelidou, K.; Morris, D. A.; Battey, N. H.; Barnett, J. R.; and John, P.; Auxin transport capacity in relation to the dwarfing effect of apple rootstocks. *J. Hortic. Sci.* **1994,** *69,* 719–725.
22. Atkinson, C. J.; Else, M. A.; Taylor, L.; and Webster, A. D.; The rootstock graft union: a contribution to the hydraulics of the worked fruit tree. *Acta Hort. (ISHS).* **2001,** *557,* 117–122.
23. Atkinson, C. J.; Else, M. A.; Taylor, L.; and Dover, C. J.; Root and stem hydraulic conductivity as determinants of growth potential in grafted trees of apple (*Malus pumila* Mill.). *J. Exp. Bot.* **2003,** *54(385),* 1221–1229.
24. Metlytskyi. Bio-Chemistry of Fruits and Vegetables. Moscow: Economics; **1970,** 271 p. (in Russian).
25. Zamorskyi, V.; Anatomic features of the young grafted apple trees. *Acta Hort. (ISHS).* **2011,** *903,* 897–902.

CHAPTER 15

METHOD FOR EVALUATION OF REGENERATION POTENTIAL OF PEAR CULTIVARS AND SPECIES (*PYRUS* L.)

ANATOLY IV. OPALKO, NATALIYA M. KUCHER, and OLGA AN. OPALKO

CONTENTS

15.1   INTRODUCTION

Pyrus L. is a member of the subtribe Pyrinae, tribe Pyreae, subfamily Spiraeoideae (formerly Maloideae), family Rosaceae. Among the widely distributed in Euro Asia, 40 species of the genus *Pyrus*, *Pyrus communis* L., and *P. pyrifolia* (Burm.) Nak. are the main edible pear species; however, *P. communis* are of the greatest economic value [1]. Most cultivars, grown in temperate zones, were created from this species. Many researchers identify intraspecific taxon—*P. communis* subsp. *domestica* (Medik.) Domin [2]—which enumerates over five thousand pear cultivars [3], and only a small percentage of them are cultivated commercially [4]. Pear cultivars grown in Japan, China, and other countries of Eastern Asia belong mostly to *P. pyrifolia* [1, 4].

About seven percent of pears are grown in commercial orchards of Ukraine, and 15 percent—in home orchards. According to this indicator, pears are second after apples; the latter occupies 75 and 40 percent of the areas allocated for fruit and berry crops, correspondingly. Last decade witnessed a positive tendency in pear production in compliance with a national program of horticulture development in Ukraine approved in 2008; it is planned to increase pear orchards to 20.8 thousand ha for the period up to the year of 2025 [5].

At present, Ukraine is among the top twenty world pear producers; its annual harvest is 102–177 thousand tons (metric tons), and it is the best index for former Soviet Union countries. Last decade showed an increasing tendency of worldwide pear production from 8 897 thousand tons in 2001 to 15 945 thousand tons in 2011. This tendency is true for Ukraine except for quite unstable pear production in various years [6]. With this in view, the requirements of quantity and quality are higher, including unification of planting pears; it also explained the necessity to do research concerning the improvement of advanced vegetative propagation of both cultivars and clonal (vegetatively propagated) pear rootstocks and their close wild relatives, which can be used in a pear-breeding program.

Pear is an allogamous plant and, as a result, economic and other features are not preserved when seed propagation is used. Thus, only vegetative propagation, including clonal pear rootstocks, can provide unified planting pears. All methods of vegetative propagation are associated with plant damage; they are based on the plants' ability to regenerate, which in turn defined the subject/direction of our research.

Various localized damages occur on the plants (especially perennial plants) during their life period. Stems of tender plants (and other organs) can be easily damaged by strong wind, frost, pouring rain, insects, different animals due to poor handling, etc. Thus, in the process of evolution, they develop an adaptive mechanism of protection against damage, that is, the ability to recover. In some cases, natural damage (also a severe one) can be a common and necessary stage of the development of an individual (plants and animals), which results in breakage, and death of some parts [7]. The ability to repair can be considered as one of the adaptive modifications; it is proportionally equal to intensity and duration of the effect of artificial and natural damaging factor. Adaptive modification takes place when intensity and duration of a damaging factor is within the limit defined by previous evolution history of a species. Nonadaptive changes, when a plant does not show adequate response and may die, can occur, provided intensity and duration of a damaging factor exceeds a standard [8].

The problem of regeneration has always been of great interest for scientists–biologists, crop growers and livestock breeders, professionals and amateurs, and in particular horticulturists [9, 10]. All signs of post-trauma regeneration can be classified into two large groups—morphogenetic regeneration, when lost parts and organs reproduce, and also a new organism from one part of a primary body can develop, including the one from a separate cell *in vitro*; and nonmorphogenetic post-trauma regeneration, which results in healing of all possible wounds [8]. The cases of so-called compensatory regeneration were described, when, after removing a part of a stem or a root, a plant reproduces a lost organ (morphogenetic regeneration), or when all the leaves but one were removed, and this only leaf grows in size considerably and provides leafless plants with photosynthesis products [11].

The results of post-trauma self-repair in plants can be explained by cambial activity [8], which, depending on phylogenetic peculiarities, is seen in basic specific and varietal distinctions concerning the ability to regenerate; both potential productivity and ecological adaptability of plants depend on it [11]. Among other factors associated with the rate and the whole course of regenerative processes, the following should be mentioned: ontogenetic peculiarities of an individual, its physiological condition, and endogenous and exogenous factors of chemical, physical, and biological nature as well. These are various chemical compounds, wound provocative, ionizing radiation, temperature and moisture of the air and soil, photo-period, phyto-sanitary condition, ontogenesis phase, and alike [8].

In the mid-thirties of the previous century, N.P. Krenke applied quantitative methods in studying age variation of somatic characteristics and formogenesis and regeneration factors, specific aspects of joining grafting components, and reasons for the formation of chimeras in plants. In his works [7], which are still actual, the importance of regeneration ability for the success of vegetative plant propagation is emphasized; this became a good foundation for further theoretical and applied research [8, 10, 11]. N.P. Krenke classified all stress factors as natural and artificial, normal and abnormal (e.g. graft of a cutting in an upside-down position), with integrity breakage of some parts or their separation from a plant, including further analysis of regeneration potential of a separated part or those parts that remained in an initial organism. The factors of grafting plant parts (structural, physiological, combined-factorial) were described, when translocation of some elements of a plant itself or introduction of unnatural elements to a plant took place, as well as other changes of plant details/parts affected by natural factors, and also bending, twisting, centrifugation, and other artificial effects [7]. Thermal, chemical, and radiation stresses can be of natural origin; they can result from by-effects of technical-genetic activity of humans or conscious special impact, etc. [8].

Plants, as well as their parts, change both quantitatively (mass, size, etc.) and qualitatively in the process of individual development. Plants move from the first (juvenile) phase, which lasts from seed germination to fruiting of a seedling, to an adult plant, and then they grow old. Such ontogenetic variation can be typical for both an individual seedling and a clone itself. Long-lived clones of grape and fruit trees keep their inheritance in most cases; however, their current properties do not always coincide with the features described by distant pomologists. Some typical physiological, anatomical, and morphological signs and properties of the earlier life period of a plant can be less seen, instead others can appear [8, 12]. Young seedlings differ from adult trees in size, leaf form, and edge serration. Young seedlings of various breeds have thorns, which are not observed on adult trees grown from these seedlings. There are also distinctions, which concern morphology and a deviation angle of laterals from a stem, productive morphogenesis, etc. In addition to the above-mentioned and many other differences, young seedlings are mostly be characterized by better regenerative-morphogenetic potentials than adult and older trees [12].

Completion of a juvenile phase of ontogenesis is usually associated with the development of floral buds, but a lower part of a young tree crown may remain in a juvenile stage, whereas flowers are formed in its upper part and a young tree enters an adult phase with an ability of fruiting [12].

The ideas of interrelationship between individual and historic development of living organisms appeared back in the XIX century, when, according to A.A. Zhuchenko, basic provisions of bio-genetic law of Friedrich Müller (1864) and Ernst Haeckel (1866) were proclaimed. According to this law, ontogenesis recapitulates major phylogenesis stages of a group, which an individual organism belongs to Zhuchenko, A.A. [13]. It is possible to regenerate roots from green cuttings of various woody plants such as coniferous, oak; most of the fruit trees and others belong to hard-rooting species, if grafting of 1- to 3-year-old seedlings is done. Most of the researchers believe that advantages in regeneration ability of green cuttings taken from young seedlings can be explained by the factor of juvenility; therefore, a seedling (in compliance with a bio-genetic law) did not totally lose the properties of ancestral forms to regenerate roots from stem parts [8, 9, 14, 15]. There is evidence that the plants, which were described in old books as those capable of adventive root formation, became hard-rooted after many years [8, 9]. At present, physiological stress called trauma/damage is considered to be an inductor of adaptive response of an organism, which facilitates regeneration [8, 16].

15.2 MATERIALS AND METHODOLOGY

Post-trauma regeneration processes in perennial woody plants and their dependence on meteorological factors were estimated according to the ability of some representatives (species and intraspecific taxons) of *Pyrus L.* to nonmorphogenetic post-trauma regeneration. Pear cultivars Bere Desiatova, Umans'ka iuvileina, Kniahynia Ol'ha and Sofiia, few rootstocks, and also cultivars of basic species *Pyrus communis* L., and several other wild relatives of pears were studied.

Pear cultivars and species studied were grown in the collections of the National dendrological park "Sofiyivka" of NAS of Ukraine, situated in the Central-Dnieper elevated region of Podilsko-Prydniprovsk area of the Forest-Steppe Zone of Ukraine. The area is characterized by temperate-continental climate with unstable humidification and considerable temperature fluctuations. Average many-year amount of precipitation per year is 633.0 mm, its amount being 300–310 mm at +10°C, which corresponds to the precipitation amount in dry southern areas of Ukraine. Average many-year air temperature is +7.4°C.

To evaluate regeneration ability, the recommendations of I.A. Bondorina [17] were used, that is, notchings (10–12 mm long, 15.5 mm wide) (Figure 15.1) with a special cutter (Figure 15.2) were made on one-year-old shoots of the previous year of the plants studied every decade from March till October. It is important to notch into the cambium.

FIGURE 15.1   The wound made with manual cutter (notching 10–12 mm long and 15.5 mm wide).

FIGURE 15.2   The manual notch cutter.

The wound, where the notching was done, was covered with transparent scotch-tape to avoid infection and withering (Figure 15.3(a)). The equation to calculate regeneration coefficient was adapted on a 9-point scale for the evaluation of regeneration efficiency [8, 10, 14]. The overgrowth of the wound was observed every decade with the help of a magnifying glass. Compare the results used of a 9-point scale. The intensity of callus genesis was estimated at 1 point, if callus formation did not occur or its surface was less than 5 percent of the wound (Figure 15.3(b)). Objects with callus areas equal to 85.5–100 percent were estimated at 9 points (Figure 15.3(c)).

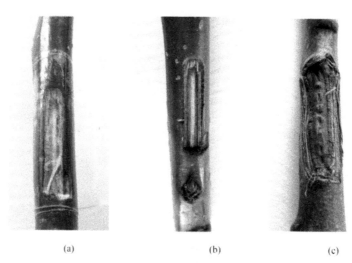

(a)   (b)   (c)

FIGURE 15.3   Estimation of the wound overgrowing: (a)—wound covered with transparent scotch-tape to avoid infection and withering; (b)—surface of callus occupies less than 5% of the wound (1 point); (c)—surface of callus occupies 85.5–100% of the wound (9 points).

Regeneration coefficient was calculated in units of regeneration coefficient (urc) according to O.A. Opalko's equation [8, 14]:

$$R = \frac{S^2}{n_1 + n_2},$$

where R—regeneration coefficient, urc;
S—intensity of callus genesis, points;
n_1—number of days after notching was done to the appearance of the first signs of callus;
n_2—number of days after notching was done to the completion or termination of callus development.

The statistical processing of the experimental data was carried out by the methods of R.A. Fisher [18]. To calculate precipitation amount and air temperature sums, the data of Uman meteorological station were used. The hydrothermal coefficient of G.G. Selyaninov [19] was calculated using the formula:

$$HTC = \frac{\sum Q}{0.1 \sum T},$$

where HTC—hydrothermal coefficient;

$\sum P$—precipitation amount;

$\sum T$—sum of temperatures higher than $+10°C$ for some period of time.

15.3 RESULTS AND DISCUSSION

The comparison of rates and intensity of wound healing to the dates of artificial notch-ings makes it possible to classify vegetative period of the cultivars and species studied based on their regeneration potentials into such stages as regeneration rise, relative decrease, a second rise wave, and rather fast damping. Besides, substantial interspe-cific and intervarietal distinctions, and in general, regeneration ability were observed, indices of regeneration coefficient fluctuating in dependence on notching terms/dates. Most likely seasonal variations of regeneration coefficient indices were due to the fluctuation of weather/meteorological conditions, namely, precipitation amount and air temperature.

In 2007–2012, average annual precipitation amount was higher in 2010 alone (by 104.1 mm). Plant moisture support in 2011 was close to average many-year indices; in 2008 and 2009 precipitation deficit was recorded (115.6 and 109.5 mm), whereas in 2007 and 2012 the deficit amounted to 217.0 and 160.1 mm, correspondingly. All deviations/variations of average daily temperatures from average many-year air tem-perature in the years studied, except for the year of 2011, tended to the increase of many-year data, the temporal variations of temperature being much smaller in a rela-tive sense than those of precipitation.

It was found out that the most favourable conditions to show regeneration potential for pear cultivars Bere Desiatova, Umans'ka iuvileina, Kniahynia Ol'ha, and Sofiia were in the year of 2010. Generalized average seasonal indices of regeneration coef-ficient in 2010 were 5.42 urc with coefficient variation over 70 percent. In 2007–2009, average regeneration potential was lower by 2.32–2.38 urc. In 2012, regeneration ex-ceeded the indices of the years of 2007–2009, but it was lower (0.66 urc) than the results of 2010. During the years of investigation the cultivars studied had higher regeneration coefficients (over 3 urc) from May till the second decade of August in-cluding, which coincided with the period of active plant vegetation. And the index of over 6 urc was recorded when notchings were made from the second decade of June till the end of the first decade of July with the maximal index 6.61 urc at the beginning of the second decade of June.

The speed at which callus developed depended on the term the notchings were made. When notchings were made in the third decade of March in all years of investi-gation, the index of regeneration coefficient of pear cultivars did not exceed two units. The most favorable period appeared to be at the end of March in 2007 (1.96 urc), whereas this indicator was only 1.32 urc in 2010. In the first decade of April, in all years of investigation, the indices of regeneration coefficient increased by 0.37–0.98 urc, and in the first half of the second decade by 0.10–0.58 urc. A slight decrease of average regeneration coefficient (0.48–0.51 urc) was recorded on the shoots damaged in the third decade of April in 2007 and 2008, whereas in 2009 and 2010 in the same period the indices of average-taxon regeneration coefficient continued to increase, and they exceeded those of the previous notching term by 0.20–0.47 urc (Table 15.1).

TABLE 15.1 Indices of pear average-taxon regeneration coefficient depending on the notching terms in 2007–2010

Notching dates	Regeneration coefficient (urc*)				
	2007	2008	2009	2010	X̄
21–23.03	1.96	1.55	1.55	1.32	1.60
01–03.04	2.45	2.53	2.49	1.69	2.29
08–14.04	3.03	3.02	2.59	2.20	2.71
19–22.04	2.52	2.54	2.79	2.67	2.63
02–04.05	2.46	3.16	2.96	3.86	3.11
12–15.05	4.67	4.63	4.49	4.32	4.53
21–23.05	5.60	5.39	5.60	4.89	5.37
01–05.06	5.33	5.03	5.36	6.67	5.60
09–14.06	4.17	4.26	4.64	7.10	5.04
21–22.06	5.58	5.73	5.69	9.42	6.61
01–04.07	4.27	4.64	4.35	12.01	6.32
08–10.07	3.96	4.36	3.84	12.79	6.24
15–20.07	4.86	4.38	4.35	8.26	5.46
25–29.07	3.93	4.35	4.25	8.22	5.19
06–09.08	2.16	2.72	2.89	10.29	4.52
16–19.08	2.04	1.96	2.69	8.72	3.85
25–31.08	1.91	1.60	1.67	2.95	2.03
07–10.09	1.24	1.16	1.17	3.70	1.82
17–22.09	1.02	0.94	1.06	1.29	1.08
26–29.09	0.63	0.69	0.58	0.68	0.65
08–14.10	0.09	0.10	0.16	0.85	0.30
X̄	3.04	3.08	3.10	5.42	3.66
S^2	1.67	1.68	1.67	3.81	-
$V_\%$	54.87	54.67	53.81	70.34	-

Note: *X—**mean;** S^2—variance; $V_\%$—coefficient of variation

In the first decade of May, 2007, average regeneration coefficient, as compared with the notchings made in the third decade of April, decreased by 0.06 urc, and in 2008–2010 increased by 0.17–1.19 urc, the smallest difference being in 2009 and the largest difference in 2010. During the second–third decades of May, the indices of regeneration coefficient continued to intensively increase in all years of investigation. In 2007, when notchings were made at the beginning of the third decade of May, the maximal/highest index of regeneration coefficient was recorded—5.60 urc, and in 2008 and 2009—5.39 and 5.60 urc, respectively. When notchings were made in the first and second decades of June, the indices of regeneration coefficient decreased slightly in 2007–2009, and at the beginning of the third decade of June the peak of the second regeneration coefficient rise was recorded. In the first decade of July, the indices of regeneration coefficient decreased gradually by 1.37–1.85 urc.

In 2010, the regeneration coefficient increased bit-by-bit, having reached its maximum 12.79 urc at the end of the first decade of July. When notchings were made in the second decade of July in 2007 and 2009, the regeneration coefficient rise was observed, its indices being 4.86 and 4.35 urc, correspondingly. In 2008 a similar regeneration coefficient rise did not take place in the same period. In 2010, in the second–third decades of July, the indices of regeneration coefficient decreased almost by 8 urc, and in the first decade of August they increased by 10.29 urc. When notchings were made in subsequent terms, regeneration coefficients gradually decreased in all years of investigation and they were zero-proximity at the end of September or at the beginning of October.

The tendency of higher dependence of regeneration coefficient on temperature fluctuation than on precipitation amount was identified in earlier researches/experiments with the representatives of *Corylus* L. [10] and *Malus* Mill [14]. This tendency was confirmed in the trials with the above-mentioned pear cultivars in 2009 and 2010. Correlation coefficients of the indices of regeneration coefficient, received in 2009 with precipitation amount (P, мм), air temperature (T, C), and hydrothermal potential (HTC), prove that higher contingency of regeneration coefficient with HTC and slight connection with precipitation amount when notchings are made (Table 15.2).

Taking into consideration the fact that in warm months when precipitation supply/support is very important, and the sums of effective temperatures decreased by one order, are nearly equal with evaporation amount, we can assume that HTC denomination in fact characterizes evaporation, and the relationship of precipitation amount and evaporation is the indicator of moisture support of the area in a vegetative season [17].

TABLE 15.2 Correlation coefficient between indices of pear regeneration ability and precipitation amount, temperature and HTC when notchings are made

Cultivar	2009			2010		
	P (мм)	T (°C)	HTC	P (мм)	T °C)	HTC
Bere Desiatova	0.14	0.51	0.49	0.35	0.82	0.50
Umans'ka iuvileina	0.26	0.51	0.73	0.46	0.73	0.38
Kniahynia Ol'ha	0.30	0.44	0.56	0.36	0.87	0.22
Sofiia	0.19	0.39	0.53	0.30	0.79	0.29

Note: P > 0.95

HTC value for regeneration increases in the years of precipitation deficit. Owing to the fact that from May till the end of August of 2009 average monthly air temperature was 0.4–4.3°C lower, and in turn evaporation was lower than in the same period of abundant precipitation in the summer of 2010, contingency of regeneration ability with HTC was higher in 2009. Precipitation amount and air temperature in 2010 exceeded the indicators in the year of 2009, which affected considerable increase of correlation coefficients between indices of regeneration ability of pear cultivars and average monthly air temperature.

In 2012, the experiment continued and another pear species *Pyrus salicifolia* Pall was added/involved. Meteorological conditions of pear vegetative season in 2012 were characterized by increased average monthly air temperature by 1.8–4.4°C, as compared with average many-year indices and monthly precipitation deficit (except for June), and accordingly by very low HTC. Under these conditions, contingency of regeneration ability of light-loving xerophyte and mesoterm *P. salicifolia* with HTC was much higher than that of long-naturalized *P. communis* (Table 15.3).

TABLE 15.3 Correlation coefficients between the indices of regeneration ability of pear species and precipitation amount, temperature and HTC when notchings are made

Species	P (мм)	T (°C)	HTC
Pyrus communis L.	0.14	0.75	0.10
Pyrus salicifolia Pall.	0.13	0.66	0.54

Note: P > 0.95

Correlation coefficients between the indices of regeneration ability of both pear species and precipitation amount were almost equally insignificant, but contingency of regeneration ability with average air temperature was rather high, *P. communis* predominating. It is worth mentioning that during the whole season 2012 regeneration ability of *P. communis* was higher that that of *P. salicifolia* (Figure 15.4). Maximal index of regeneration coefficient of *P. communis* amounted to 10.12 urc in the variant when notching was made on August 20, whereas the best indicator of *P. salicifolia* hardly reached 2.45 urc (the notching was made on July 11). All this gives every ground to assume that the indicator of regeneration ability can prove the level of ecological adaptation of the genotypes studied.

With the exception of the third decade of March (the beginning of the experiment) and the third decade of September (the fast damping of regeneration rates), when the regenerative potential of *P. salicifolia* was slightly higher than *P. communis*, the curves of a seasonal variation of the both species for the *year 2012* were quite similar.

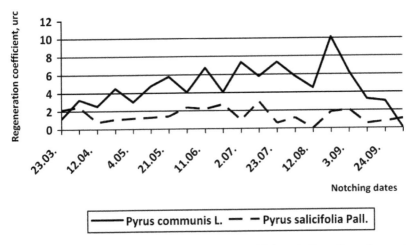

FIGURE 15.4 Seasonal variations of regeneration coefficients indices *Pyrus* L. species in 2012

The other indices were in 2013. In early May and late August, the excesses were observed in *P. communis* and at the end of the first decade of July the unexpected excess was observed in *P. salicifolia* (Figure 15.5).

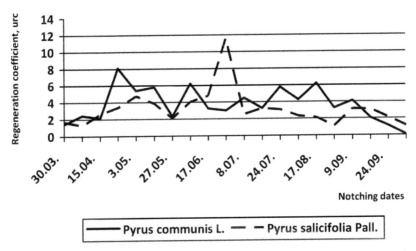

FIGURE 15.5 Seasonal variations of regeneration coefficients indices *Pyrus* L. species in 2013.

The interspecies differences in the indices of a regenerative ability on the other dates of season 2013 were insignificant or statistically unproved. It has motivated for the comparison of pear average-taxon regeneration coefficients depending on the notching terms in 2012 and 2013.

The curves of variation of indices of regenerative coefficients in 2012 do not have great excesses with the exception of the first decade of September, when the regenerative potential of *P. communis* was maximum for the season and *P. salicifolia* had a slightly higher than average seasonal indices. The variation of indices of regenerative coefficients in 2013 was considerably higher, that is, due to the instability of weather conditions in 2013 (Figure 15.6).

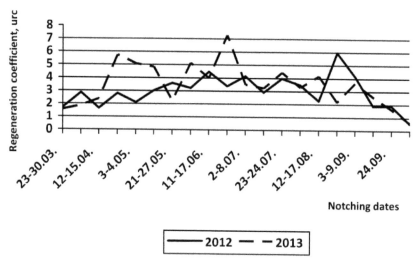

FIGURE 15.6 Indices of pear average-taxon regeneration coefficients depending on the notching terms in 2012 and 2013.

15.4 CONCLUSION

The intensity of nonmorphogenetic post-trauma regeneration in *Pyrus* species studied was not the same and changed in the years of investigation: it depended more on temperature fluctuation than on precipitation amount and hydrothermal coefficient. The assumption can be made that periods of the highest regenerative activity can be favorable for cutting, grafting, micro-cloning, etc., as well as for various technological processes, associated with trauma/damage.

ACKNOWLEDGMENTS

This study is partly based on the work supported by the National dendrological park "Sofiyivka" of NAS of Ukraine (No 0106U009045) together with Uman National University of Horticulture (No 0101U004495) in compliance with their thematic plans of the research work. We thank corresponding member of NAS of Ukraine Ivan Kosenko, PhDs. Michael Nebykov and Aleksandr Serzhuk as well as graduate student Vitaliy Adamenko for their help with scientific-technical programs and for consultations and discussion.

KEYWORDS

- **Adaptive modifications**
- **Clonal pear rootstocks**
- **Hydrothermal coefficient**
- **In vitro**
- **Juvenility**
- **Notching**
- **Ontogenesis**
- **Physiological stress**
- **Regeneration**
- **Vegetative propagation**

REFERENCES

1. Bell, R. L.; and Itai, A.; Pyrus. Wild Crop Relatives: Genomic and Breeding Resources, Temperate Fruits. Ed. Chittaranjan Kole. Berlin: Heidelberg: Springer; **2011,** *8,* 147–178.
2. Catalogue of Life: 2010 Annual Checklist. Catalogue by Royal Botanical Gardens Kew. URL: http://www.catalogueoflife.org/annual-checklist/2010/details/species/id/7179820 (Accessed June 12, 2013).
3. Yamamoto, T.; and Chevreau, E.; Pear Genomics. Genetics and Genomics of Rosaceae, Plant Genetics and Genomics: Crops and Models. Ed. Folta, Kevin M.; and Gardiner, Susan E.; New York: Springer Science + Business Media, LLC; **2009,** *6,* 8. 163–188.
4. Monte-Corvo, L.; Goulao, L.; and Oliveira, C.; ISSR analysis of cultivars of pear and suitability of molecular markers for clone discrimination. *J. Am. Soc. Hortic. Sci.* **2001,** *126(5),* 517–522.
5. Horticulture Development Sectoral Programme of Ukraine for the Period Until 2025. Kyiv: Zhytelev; **2008,** 76 p. (in Ukrainian).
6. Pear Production. Food and Agriculture Organization of the United Nations. **2011,** FAOSTAT. URL: http://faostat.fao.org/site/567/DesktopDefault.aspx?PageID=567#ancor
7. Krenke, N. P.; Regeneration of Plants. Moscow; Leningrad: Academy of Sciences of the USSR Press; **1950,** 667 p. (in Russian).
8. Kosenko, I. S.; Opalko, O. A.; and Opalko, A. I.; Posttravmatic regeneration processes at plants. Autochthonous and alien plants. The collection of proceedings of the National dendrological park "Sofiyivka" of NAS of Ukraine. **2008,** *3–4,* 10–15 (in Ukrainian).
9. Hartmann, H. T.; and Kester, D. E.; Plant Propagation: Principles and Practices. Englewood Cliffs: Prentice-Hall Inc; **1975,** 622 p.
10. Kosenko, I. S.; Opalko, A. I.; and Sergienko, N. V.; Posttraumatic regeneration processes of the representatives of *Corylus* L. genus. Proceedings of the Russian Scientific conference with international participation "Botanical gardens in the developing world: theoretical and applied research" (July 5–7, 2011, The Tsytsin Main Moscow Botanical Garden of RAS). Ed. Aleksandr Demidov. Moscow: KMK Scientific Press Ltd; **2011,** 347–350 (in Russian).
11. Jusufov, A. G.; On the Evolution of the Phenomena of Regeneration of Plants. Makhachkala: The Science of Plant Regeneration; **1978,** 53–55 (in Russian).
12. Visser, T.; Juvenile Phase and Growth of Apple and Pear Seedlings. Euphytica; **1964,** *13(2),* 119–129.
13. Zhuchenko, A. A.; Adaptive Potential of Cultivated Plants (Genetic and Ecological Bases). Kishinev: Shtiinca; **1988,** 768 p. (in Russian).

14. Opalko, O. A.; and Opalko, A. I.; Regenerative capacity as criterion of the use of representatives of the genus *Malus* Mill. in landscape compositions. *Proc. Tbilisi Bot. Garden.* **2006,** *96,* 187–189 (in Russian).

15. Sinnott, E. W.; Plant Morphogenesis. Whitefish: Literary Licensing, LLC; **2012,** 560 p.

16. Opalko, O.; and Balabak, O.; Physiological Stress—Rhizogenic Inductor Activity of Garden Plants Cuttings. Bulletin of the Lviv State Agrarian University: Agriculture; **1999,** *4,* 179–181 p. (in Ukrainian).

17. Bondorina, I. A.; Principles for Improved the Properties of Ornamental Woody Plants Grafting Techniques: Abstract of Thesis of Candidate of Biological Sciences: 03.00.05. Moscow: The Tsytsin Main Moscow Botanical Garden of RAS; **2000,** 21 p. (in Russian).

18. Fisher, R. A.; Statistical Methods for Research Workers. New Delhi: Cosmo Publications; **2006,** 354 p.

19. Shein, E. V.; and Goncharov, V. M.; Agrophysics. Moscow: Feniks; **2006,** 400 p. (in Russian).

CHAPTER 16

GENETIC RESOURCES OF THE GENUS *CORYLUS* L. IN THE NATIONAL DENDROLOGICAL PARK "SOFIYIVKA" OF NAS OF UKRAINE

IVAN SEM. KOSENKO

CONTENTS

16.1 INTRODUCTION

Since ancient times, the nuts of cultivated and wild representatives of the genus *Corylus* L. are used with the aim of food. It is explained with the high nutritive and dietary value of the kernels, as they consist of 45–47 percent of lipids, where oleic and palmitic acids dominate (85 and 10%), 12–25 percent proteins, sugar, vitamins, including 40–50 mg per 100 g of tocopherol (vitamin E) [1, 2].

The history of hazelnut versatile use is full of secrecy and unearthly effects. As we all know, in the culture of Ancient Greece and Turkey hazelnut is the symbol of peace. According to the myths and legends of an Ancient Greece, the warder of the God of commerce Hermes was made from hazelnut. Since that time, the all known "magic wand" or "magic staff," which was used by treasure divers, were produced from hazelnut. Hazelnut was dispensed with the magical properties due to the belief of many nations. Arabs, for instance, believed that the wand made of hazelnut protect their owners from all the viciousness. The people from the north Europe supposed that hazelnut protects the home from lightning and drives off the witches. Besides, hazelnuts were used as the symbol of fertility in wedding ceremonies, and the guarantee of fortunate marriage and numerous physically health posterity. Vergili glorified the hazelnut as the plant that deserves a lot of estimation, even more than "grapes or laurel" [2].

There is only one native species of hazelnut, *Corylus avellana*, in the natural flora of Ukraine. We do not know exactly when the hazelnut appeared; the opinions of some scientists differ and dated from the beginning of neogen–paleogen in the Cainozoic era till the end of Cretaceous period of Mesozoic era. Anyway, after the pleystotsen icing-over (approximately at the eight-five millenary before BC), hazelnut dominated in the natural flora of the northern Europe [1].

At the end of between ice age, because of the beginning of the fall in temperature, plants and animals perished in mass, thus *Corylus avellana* completely disappeared from the forests of the northern Europe. However, the life was saved at the some ecological niches, and seclusion center. Hazelnut also was saved and distributed in the future over the rest of the useful territory. The mass distribution of the hazelnut *C. avellana* over Europe started after the subarctic time when the pine tree dominated in the forests. Hazelnut moved to the north together with the birch, alder tree, oak tree, and pine tree. Under the favorable climate conditions, the hazelnuts under the tent of pine tree created such thick brushwood that it prevented the natural renewal of the other plants, such as pine tree and birch tree. Owing to it, at the place of pine and birch forests, hazelnut groves appeared. The fir tree became an essential obstacle for the further distribution of hazelnut. However, the loose dominating positions in the general vegetation did not restrict the natural habitat of the hazelnut [1, 2].

Together with the development of intense horticulture, hazelnut gained importance as a nut fruit crops. Moreover, hazelnut won its place in the ornamental horticulture. Nowadays, the species *C. colurna* and *C. maxima Mill.* are widely used in the planting of greenery of different towns; the rest of species are kept at the collections of botanical gardens, dendroparks, and reserves. However, the experience of cultivation of the most of the valuable forms of the northern American and Eastern Asia species testifies the great prospects for the hazelnut establishment in the ornamental horticulture [1].

Some species and forms of the hazelnut could be used for slope fortification and in the other plantings, because of the hazelnut resistance to unfavorable factors of an environment [1].

During the long period, hazelnuts were stored up at the wild forests. For the first time, cultural filbert cultivars appeared in Turkey, more than 2 thousand years ago, that is why the filbert culture in this country is well developed. At the beginning of another century, the filbert production in the world market exceeded 800 thousand (metric) tons. Turkey occupied 70 percent of filbert production with the annual manufacture of 450–600 thousand tons of filberts. Italy takes the second place steady with the index 100–130 thousand tons; the third place is for the United States—15–40 thousand tons. The next is Spain—15–25 thousand tons, Azerbaijan, Iran, Greece, and China produce 6–15 thousand tons per year, and France—4–5 thousand tons of filbert nuts per year. Nearly all these countries are absolutely self-sufficient and able to provide its interior markets. They sell a lot of filbert nuts on the outside markets, but at the same time Italy buys 30–35, France—nearly 20, and Spain—nearly 1 thousand tons of hazelnut. Nearly 75 percent of world production of husked nuts is used for chocolate industry and only 25 percent for the other confectionary goods, pastas, and admixtures to the grainy and fruit mixtures the so-called "dry breakfasts" [2].

At present, the general area with filbert plantings in the economies of all categories and forms of Ukraine so far does not exceed 0.1 thousand ha. Filbert productivity at this plantings is on the average 0.18–0.43 t/ha, including the territory of agro-industrial enterprises—0.01–0.13, and in the personal garden land of the population—1.10–3.15 t/ha. The gross yield of filbert nuts does not exceed 20–40 tons a year. Moreover, certain amount of hazelnuts are gathered, while the official statistics for the gross yield of hazelnut is not conducted. Observing the state of chocolate and confectionary interior market, we may confirm that this number is not enough in order to provide even inside requirements. In spite of that, the soil and climatic conditions of Ukraine give the opportunity to grow 1.0–1.5 t/ha filbert nuts with a profitability of more than 300 percent; the nuts and nut production are still imported in to Ukraine from the foreign countries [1, 2]. Nearly 2.8–3.1 thousand tons of filbert nuts on the sum US\$4.4–5.5 million are imported into Ukraine every year. The third part of it is purchased in Turkey and the rest in Azerbaijan, Georgia, and other European countries [2].

Filbert is the collective name of all cultivated *Corylus* species (hazelnut cultivars). At present, most of the investigators suppose that the largest part of filbert heredity came from the hazelnut *Corylus avellana*, so also for the cultivars that like warm came from *Corylus pontica* and *Corylus maxima*. *C. colurna* was used for creation of tree-like and sobole-free filbert cultivars [1–3].

16.2 MATERIALS AND METHODOLOGY

Specific and variety-form-hybrid composition of the collection of *Corylus* genus at the National dendrological park "Sofiyivka" of NAS of Ukraine (NDP "Sofiyivka") was studied. Most of the species and forms of this collection were gathered during the numerous expeditions over a period of 80–90 years of the last century, whereas the maximum filling of this sort of variety was done in the course of the last decade. On November 7, 2006, from the collection of the Institute of Horticulture of the National

Academy of Agrarian Sciences of Ukraine, the large-sized cultivars of filberts were dug out and planted at the newly created "Filbert Nursery." According to the universally adopted methods, the morphology of represented species, sorts, and forms were investigated [1, 4].

NDP "Sofiyivka" is situated in the town of Uman, Cherkasy region. The park is well known, though far away in the borders of Ukraine, as one of the most famous masterpieces of the world landscape gardening of the end of XYIII—the beginning of XIX centuries. "Sofiyivka" stands at the same level with other examples of park construction of Europe, such as Boboli Garden in Florence (Italy), San-Susi Park in Potsdam (Germany), and Versailles (France) [1].

Its geographical coordinates are 48°46' of Northern latitude and 30°14' of Eastern longitude by Greenwich. The height above the sea level fluctuates in the diver's places of the park from 170 to 216 m. Nowadays, the area of the park is over 180 ha (Figure 16.1).

The park stands on the granite bazolit of mazogmatic character. The granite is nearly 20–40 m depth, but in the gullies and channels appears on the surface and makes picturesque cliffs. The main components of the soil are loess and the modern alluvial and delluvial bottom measures of gullies and valleys.

The territory of the park is situated in the valley of the Kamianka River, which crosses the park in the southwestern direction through the valley and interblock of the Grekiv ravine and Menagerie.

During the period of park's creation, the damp and the cascade of four ponds were created along the channel of the Kamianka River. This cascade became the new element of the park's relief and influenced greatly on the microclimate and processes of new creations [1].

The climate of the region is moderate. The annual average temperature is 7.3°C. The total temperature during the period with the temperature above 10°C is 2,500–2,800. In the average temperature of June, the hottest month is 19°C; the highest temperature is 38°C. The average temperature of January, the coldest month is −4.4°C; the lowest temperature was noted −37°C.

The annual rainfall changes from 339 to 949 mm, and on average it is nearly 633 mm. In winter, the ground freezes nearly to 75–80 sm and sometimes even to 120 sm deep. The blanket of snow 15–18 cm thick appears in the middle of December and stays on the surface for 120 days. But it is not steady as there are frequent thaws when the temperature rises to 9–12°C. Usually the ground melts absolutely in the third decade of March. The frosts are also possible early in the autumn and at the end of May. The climadiagram of the region of hazelnut introduction is shown in Figure 16.2.

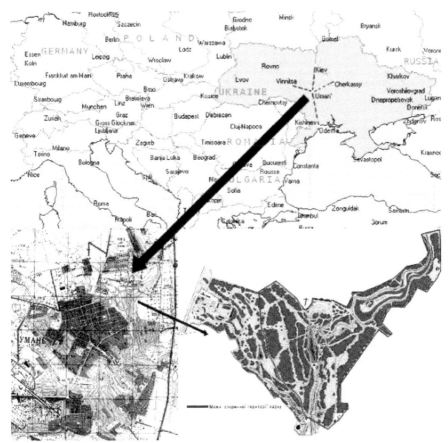

FIGURE 16.1   The geographical position and schematic sketch of the National dendrological park "Sofiyivka".

There are various soils on the territory of the park, they are regraded chernozems, taupe forest ashed soils, taupe and lightly washed forest soils, meadow and swampy soils, and soils at conceptual stage of forming. From the botanical point of view, the park is situated in the Right bank of western Forest Steppe Middle-Dnieper region of CIS flora, and the geographical, climatic, and orthographical park conditions create the necessary conditions for the significant growth of the forest, meadow, boggy, and hydrophyte species of plants [1].

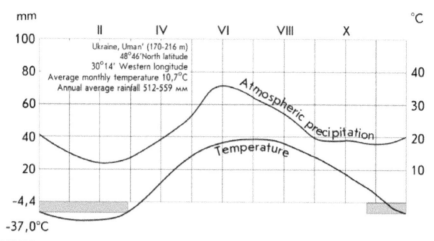

FIGURE 16.2   The climadiagram of the region of hazelnut introduction.

Nowadays, "Sofiyivka" is not only the "Historical garden" but also the scientific research institution of the National academy of Science of Ukraine, the center of introduction, mobilization, and acclimatization of plant biodiversity in the Right bank Forest Steppe zone of Ukraine, the teaching and educational base, the tourist's institution, and the museum of landscaping art. Nearly 3,500 species, forms, sorts, and hybrids of trees and grassy plants are collected here, including the collection of the genus *Corylus* L. [5–7].

16.3   RESULTS AND DISCUSSION

The genus *Corylus* L. (hazelnut) is a member of the family *Betulaceae* S.F. Gray, subfamily *Coryloideae* [8]. Some time earlier, a lot of authors considered it reasonable to mark out *Coryloideae* as an independent family *Corylaceae* Mirb. [9, 10], and we also supported this opinion [1], taking into consideration the morphological marks and well-known suggestion as for the removal of genera *Carpinus*, *Corylus*, and *Ostrya* from *Betulaceae*, so that to consolidate them in the family *Corylaceae* [11]. Scientists from Cornell University (USA) and Royal Botanical Gardens Kew (Great Britain) wrote a common monograph and described botanical names and synonyms of tribes and genera *Magnoliidae*; the name *Corylaceae* is given as one of the synonyms of *Betulaceae* [12].

At the joint catalogue The Plant List by the Royal Botanical Gardens Kew and Missouri Botanical by 2010 [13], 138 Latin names for the representatives of the genus *Corylus* are indicated: 26 were recognized (17 specific and 9 interspecific taxons). On July 2011, 134 names (74 species and 60 interspecific taxons), so as 27 recognized names (17 specific and 10 interspecific taxons) were included in the Catalogue Royal Botanical Gardens Kew, Richmond, Surrey [14].

At the beginning of 2011, the collection of NDP "Sofiyivka" counted 33 taxons of the genus *Corylus* [15], 17 of which are in the class of species. Having analyzed the collection according to the morphological features, we cleared up that only 11 species should be marked. All the others should be considered as synonyms, interspecific tax-ons (Table 16.1), and alien cultivars and forms.

TABLE 16.1   The structure of the collection of genus *Corylus*

S. No.	Collection of NDP "Sofiyivka" of NAS of Ukraine	Catalogue Royal Botanical Gardens, Kew Richmond Surrey (UK)	The Plant List by the Royal Botanic Gardens, Kew (UK) and Missouri Botanical Garden (USA)
1	*C. abchasica* Kem.-Nath.	Synonym *C. colurna* L.	Synonym *C. colurna*
2	*C. americana* Walter	*C. americana*	*C. americana*
3	*C. avellana* L.	*C. avellana*	*C. avellana*
4	*C. pontica* (K. Koch) H.J.P. Winkl.	Synonym *C. avellana* var. *pontica*	Synonym *C. avellana* var. *pontica*
5	*C. chinensis* Franch.	*C. chinensis*	*C. chinensis*
6	*C. colchica* Albov	*C. colchica*	*C. colchica*
7	*C. colurna* L.	*C. colurna*	*C. colurna*
8	*C. colurnoides* C.K.Schneid.	*C. colurnoides*	*C. colurnoides*
9	**C.** *cornuta* **Marshall**	**C.** *cornuta*	C. cornuta
10	*C. domestica* Kos. et Opal.	—	—
11	*C. tibetica* (Batalin) Franch.	Synonym *C. ferox* var. *tibetica*	Synonym *C. ferox* var. *tibetica*
12	*C. heterophylla* Fisch. ex Trautv.	*C. heterophylla*	*C. heterophylla*
15	*C. iberica* Wittm. ex Kem.-Nath.	Synonym *C. colchica* Albov	Synonym *C. colchica*
14	*C. jacquemontii* Decne.	*C. jacquemontii*	*C. jacquemontii*
15	*C. maxima* Mill.	*C.maxima*	*C. maxima*
16	*C. sieboldiana* Blume	*C. sieboldiana*	*C. sieboldiana*
17	*C. mandshurica* (Maxim.) C.K.Schneid.	Synonym *C. sieboldiana* var. *Mandshurica*	Synonym *C. sieboldiana* var. *mandshurica*

Note: half-thick type marked with recognized specific names

Since the ancient times, the populations' *C. avellana*, *C. Maxima*, and *C. avellana* var. *pontica* (*C. Pontica* C. Koch) have been growing as nut trees. Considerably later, *C. americana* was introduced into the culture, though in the cultivar variety of this species we discovered the forms that according to the size and form of nuts were similar to the representatives *C. avellana* L., *C. Maxima* Mill. and *C. avellana* var. *pontica*. The following cross between them, such as with *C. colurna*, *C. americana* and other species of the genus *Corylus*, provided the possibility of obtaining a lot of modern filbert cultivars. Hereby, there is no saying about the part of heredity for each species in the genotype formation [1–3, 16–19]. That is why, at the International scientific conference in Kyiv (Ukraine), October 2007; we suggested to combine all filbert cultivars into one collective species *C. domestica* Kos. et al. [20].

State Services for Plant Variety Rights Protection included at the state register of cultivars, which are suitable for distribution in Ukraine in 2011 only four filbert cultivars, such as 'Ata-baba,' 'Barcelona,' 'Galle,' and 'Kosford' [21]. Hereby, all were assembled under the general name big Hazelnut (filbert), in Latin *Corylus maxima* Mill. and in English Red Filbert, which is not right. As a big Hazelnut, it is just a word for word translation from Latin for *Corylus maxima*, and 'Red Filbert' is used in a great many of English sources as the name of filbert cultivar that belong to *Corylus avellana* Zellernus (unresolved name) with the mark that this cultivar is famous as *Corylus maxima* 'Purpurea.' At the great catalogue of woody plants [22], which was issued in the North Caroline (USA) Red Filbert, is described as a filbert cultivar that belong to *Corylus maxima*, and at the same time, the synonyms of this sort 'Rote Zeller,' 'Red Zellernut,' and 'Rote Zellernus' are indicated; it is mentioned that this cultivar, which was created in Germany, is distributed by the nurseries of the Royal British Horticulture Society as the representative of *Corylus avellana*. Such a discrepancy in the classification affirms the necessity of its improvement, that is why we suggest to establish generally the species name we proposed [2, 20], *C. domestica* Kos. et Opal.

As a result of the material generalization, which were done using the analysis of DNA-sequence [18–23], together with our investigations, we ascertained that some species such as *C. abchasica* and *C. iberica*, which we declared in our earlier publications [1, 24], exposed to be the synonyms to *C. colurna* L. and *C. colchica* suitably; then the species *C. pontica*, *C. tibetica* and *C. mandshurica* are varieties of *C. avellana* var. *pontica*, *C. ferox* var. *tibetica*, and *C. sieboldiana* var. *mandshurica*.

All the data, which were obtained in the result of DNA-sequence analyzed [18, 23], were used by them for the definition of an interspecific link in the genus *Corylus* (Figure 16.3). It was suggested to embody all the studied species into the two sections:

- *Acanthochlamys* with the base species *C. ferox*;
- *Corylus*, which unite the rest of the species of the genus;

In this section three subsections were recognized:

- *Corylus* (shrub species with foliage, well-developed, with the cover of fruits that resemble bluebells);
- *Colurnae* (the species of vital form 'tree' and with deeply dismembered fruit cover);
- *Siphonochlamys* (shrub species with tubular fruit cover).

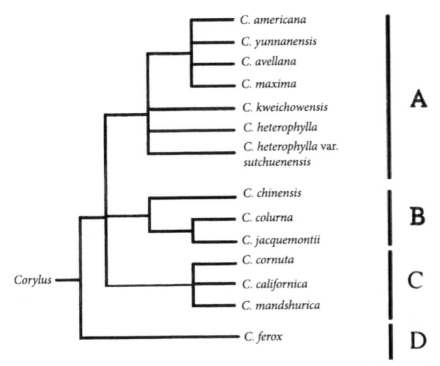

FIGURE 16.3   The principle scheme of interspecific relations in the genus *Corylus* (according to Whitcher I.N. and Wan J., 2001 [23] with changes): A—section *Corylus*, subsection *Corylus*, fruit cover is well-developed; B—section *Corylus*, subsection *Colurnae*, vital form—tree; C—section *Corylus*, subsection *Siphonochlamys*, tubular fruit cover; D—section *Acanthochlamys*, fruit cover is covered with thorns.

In our opinion, this cladograma reflects the interspecific links in the genus *Corylus* fairly well, even if it does not include all the recognized species. Comparative analysis of species with the species names of the genus *Corylus* collections in the databases of leading botanical institutions [13], which are indicated here confirms the sufficient representative (as for the species variety) level of cited investigations [23]. At the same time, *C. americana*, *C. avellana*, *C. Maxima*, and *C. yunnanensis* were united into the common group, which is very close to the philogenetics elder group, which include *C. heterophylla* and *C. heterophylla* var. *sutchuenensis* with the definition that *Corylus kweichowensis* is supposed to be the synonym of *Corylus heterophylla* var. *sutchuenensis* Franch. All together, they form the subsection *Corylus* in the section *Corylus* with the shrub species and foliage, well-developed fruit cover, which resembles a bluebell. The species of the vital form 'tree' *C. chinensis*, *C. Colurna*, and *C. jacquemontii* form the subsection *Colurnae* with the basic species *C. chinensis*. Subsection *Siphonochlamys* has typical tubular fruit cover is presented with monophilic group *C. cornuta* with definitions that *C. californica* (A.DC.) A. Heller is the synonym of *C. cornuta* subsp. *californica* (A.DC.) A.E. Murray [13]. The monophilic species is *C. ferox* with tree-like plants and fruits, the cover of which is filled with thorns belong to the section *Acanthochlamys*.

The representatives of some recognized species are absent in the collection of NDP 'Sofiyivka,' they are *C. fargesii* (Franch.), *C. ferox* Wall., *C. wangii* Hu, *C. wulingensis* Q.X. Liu & C.M. Zhang, *C. yunnanensis* (Franch.) A. Camus, and *C. potaninii* Bobrov. We will find the five of them, whereas we doubt the existence of *C. potaninii*.

In spite of the fact that in some publications [25], such as in the database of some leading institutions [13, 14], *C. iberica* is supposed to be the synonym of *C. colchica* Albov, we are sure that *C. iberica* is the synonym of *C. colurna* instead of *C. colchica*. Referring to Karl Vitman, he was the first person who had named, but had not described, *C. iberica*. L.M. Kemularia Natadze describes it as an independent dendritic species with Caucasian areal [26]. The plants *C. iberica* were grown from the seeds that we gathered in the Zakatalski reserve (Azerbaijan) and in the Tbilisi Botanical Garden, and did not have considerable differences from the typical trees *C. colurna*. However, *C. colchica* Albov has the vital form of a small bush, 0.5–1.0 m high, and does not look like the plants of *C. iberica* [1].

Having enriched the own collection essentially with the filbert cultivars from the Institute of Horticulture of NAAS of Ukraine, with the materials from the collection of the Pomology Institute nd. a. L.P. Symyrenko of NAAS of Ukraine and nursery gardens of Poland—nowadays the collection of filbert cultivars in the NDP 'Sofiyvika' consists of 132 examples, 90 of which are the cultivars of domestic and foreign breeding, and 42 breeding numbers. Most of the domestic cultivars and breeding numbers were created by the famous breeder F.A. Pavlenko, who started his breeding work with hazelnut in 1938 under the guidance of Prof. S.S. Piatnitskiy and acheived a lot of success [27, 28].

The representatives of ornamental forms of the genus *Corylus* are included in the collection of NDP 'Sofiyivika,' they are *C. avellana* (*C. a.* f. *contorta*, *C. a.* f. *contorta* 'Red Majestic,' *C. a.* f. *fuscorubra* Dippel., *C. a.* f. *laciniata* Petz et Kirchn., *C. a.* f. *pendula* Goeschke, *C. a.* f. *praecocs*, *C. a.* 'Red Filazel'), *C. colurna* (*C. c.* f. *globosa*, *C. c.* f. *intermedia*, *C. c.* 'Nadia,' *C. c.* 'Poltavska,' *C. c.* f. *praecocs*, *C. c.* f. *pyramidalis*, *C. c.* f. *serotina*), and *C. maxima* (*C. m.* f. *atropurpurea* Dochnahl).

16.4   CONCLUSION

Nowadays, the collection of the genus *Corylus* in the NDP 'Sofiyivka' is the largest fund of genetic resources in Ukraine that can be used as source material for breeding, as well as as an ovary for propagation by layers (species and cultivars) and by grafting on stem of *C. colurna* for ornamental forms, which are very popular in the market of seedlings.

ACKNOWLEDGMENTS

This material is partly based on the work supported by the National dendrological park 'Sofiyivka' of NAS of Ukraine together with the Institute of Horticulture of the National Academy of Agrarian Sciences of Ukraine, the Pomology Institute nd. a. L. P. Symyrenko of the National Academy of Agrarian Sciences of Ukraine, and the

Dendrological Park of National Assignment "Veseli Bokovenky" nd. a. M.L. Davydova. We are thankful for their help with scientific-technical programs and for consultations and discussion with Associate professor Ph.D. Anatoly Opalko, Ph.D., Oleksandr Balabak, and Junior Researcher Galina Tarasenko.

KEYWORDS

- **Alien species**
- **Breeding cultivar**
- **Collection**
- **Filbert**
- **Hazelnut**
- **Introduction**
- **Sobole**

REFERENCES

1. Kosenko, I. S.; Hazelnuts in Ukraine. Ed. Kokhno, M. A.; Kyiv: Academperiodyka; **2002**, 266 p. (in Ukrainian).
2. Kosenko, I. S.; Opalko, A. I.; and Opalko, O. A.; Filbert: Applied Genetics, Breeding, the Methods of Propagation and Production. Ed. Kosenko, I. S.; Kyiv: Naukova Dumka; **2008**, 256 p. (in Ukrainian).
3. Opalko, A. I.; Nut plants breeding. In: Opalko, A. I.; and Zaplichko, F. O.; Fruit and Vegetable Breeding. Kyiv: Higher School; **2000**, 386–398, 440 p. (in Ukrainian).
4. Methodics of Phenological Observations in the Botanical Gardens of USSR. Moscow: Nauka; **1975**, 23 p. (in Russian).
5. Kosenko, I.; Collection funds of the genus *Corylus* L. in the National Dendrological Park "Sofiyivka" as a valuable base for filbert breeding (VI International Congress on Hazelnut). *Acta Hort. (ISHS).* **2005**, *686*, 587–602.
6. Kosenko, I. S.; Opalko, A. I.; and Tarasenko, H. A.; Genetic resources of the *Corylus* L. genus to Ukraine. IHC Lisboa; **2010**, 571 p.
7. Kosenko, I. S.; The genetic resources mobilization of *Corylus* L. genus at the National Dendrological Park "Sofiyivka" of the NAS of Ukraine. News Biosphere Reserve. Askania Nova; **2012**, *14*, 156–160. (in Ukrainian).
8. Takhtajan, A. L.; Flowering Plants. Corr. 2nd edition. New York: Springer Science + Business Media; **2009**, 871 p.
9. Hutchinson, J.; and Bentham, G.; The genera of flowering plants (*Angiospermae*). Oxford: Clarendon Press; **1964**, *1*, Dicotyledones. 516 p.
10. Dahlgren, R.; General aspects of angiosperm evolution and macrosystematics. *Nordic J. Bot.* **1983**, *3(1)*, 119–149.
11. Mosyakin, S. L.; and Fedoronchuk, M. M.; Vascular plants of Ukraine. A Nomenclatural Checklist. Kiev: M.G. Kholodny Institute of Botany; **1999**, 345 p.
12. Reveal, J. L.; and Chase, M. W.; APG III: Bibliographical Information and Synonymy of *Magnoliidae*. Phytotaxa; **2011**, *19*, 71–134.
13. The Plant List by the Royal Botanic Gardens Kew and Missouri Botanical **2010**, URL: http://www.theplantlist.org/tpl/search?q=Corylus
14. Catalogue of Life: 26th July 2011 Catalogue by Royal Botanical Gardens Kew **2011**, URL: http://www.catalogueoflife.org/testcol/search/all/key/Corylus/match/1

15. Kosenko, I. S.; Hrabovyi, V. M.; Moroz, O. K.; et al., Dynamics of the taxonomic structure of plantings in the National dendrological park "Sofiyivka" of NAS of Ukraine. Autochthonous and alien plants. The collection of proceedings of the National dendrological park "Sofiyivka" of NAS of Ukraine. **2011**, *7*, 15–24. (in Ukrainian).

16. Bobrov, E. G.; The history and systematic of the genus *Corylus*. *Soviet Bot.* **1936**, *1*, 11–39. (in Russian).

17. Brunner, F.; and Fairbrothers, D. E.; Serological investigation of the *Corylaceae*. *Bull. Torrey Botanical Club.* **1979**, *106(2)*, 97–103.

18. Erdogan, V.; and Mehlenbacher, S. A.; Interspecific hybridisation in hazelnut (*Corylus*). *J. Amer. Soc. Hort. Sci.* **2000**, *125(4)*, 489–497.

19. Thompson, M. M.; Lagerstedt, H. B.; and Mehlenbacher, S. A.; Hazelnuts. Fruit breeding. Ed. Janick, Nuts J.; and Moore, J.; John Wiley and Sons, Inc; **1996**, *III*, 125–184.

20. Kosenko, I. S.; and Opalko, A. I.; Dynamics of *Corylus* L. genus as corroboration of N. I. Vavilov's law about the homologous rows in the inheritable variability. In: Introduction of Plants at the Beginning of the XXI Century: Achievement and Perspectives. Proc. of the II Inter. Sc. Conf. devoted to the 120th anniversary from the day of birth of academician. Vavilov, N. I.; Kyiv: Phytosotiocentre; **2007**, 70–74. (in Ukrainian).

21. State register of cultivars which are suitable for distribution in Ukraine in **2011**, (from 05.09.2011) URL: http://sops.gov.ua/uploads/files/documents/reyestr_sort/R2011_05.09.11.pdf

22. Hatch, L. C.; Cultivars of Woody Plants. Raleigh: TCR Press; **2007**, *I(A–G)*, 1031 p.

23. Whitcher, I. N.; and Wen, J.; Phylogeny and biogeography of *Corylus* (*Betulaceae*): Inferences from ITS sequences. *Syst Botany.* **2001**, *26(2)*, 283–298.

24. Bilyk, O. V.; Vegera, L. V.; Dzhym, M. M.; et al., Plant catalogue of the dendrological park "Sofiyivka" of NAS of Ukraine. Ed. Kosenko, I. S.; Uman: NDP "Sofiyivka" of NAS of Ukraine; **2000**, 160 p. (in Ukrainian).

25. Cherepanov, S. K.; Vascular Plants of Russia and Allied States. St. Petersburg: World and Family; **1995**, 990 p. (in Russian).

26. Kemularia Natadze, L. M.; Treelike hazelnut in Georgia and its hybrids. Works of Tbilisi Botanical Institute; **1938**, T. *6*, 1–24 p. (in Russian).

27. Pavlenko, F. A.; Slyusarchuk, V. E.; and Sytnik, I. I.; Industrial Filbert Propagation: Survey Information. Moscow: CBSTI Gosleshoz USSR; **1988**, Edition 1, 36 p. (in Russian).

28. Slyusarchuk, V. E.; Biodiversity of Hazelnuts and Filberts: Conservation and Enrichment. Scientific Herald of Ukraine: The Collective Materials of Scientific and Technical Works. **2006**, Edition 16.6, 11–18.

CHAPTER 17

SOIL MULCHING AFFECTS FRUIT QUALITY IN STRAWBERRY

IRINA L. ZAMORSKA

CONTENTS

17.1   INTRODUCTION

Strawberry is one of the most popular fruit crops in the world, which is due to its gustable, curative-dietary properties, and high economic efficiency of cultivation. Strawberries are an excellent dessert product owing to their tender pulp, easy assimilability, and a balance of sugars and acids of the berries.

Berries are consumed as fresh and preserves. Strawberry plants are rich in sugars, organic acids, microelements, and vitamins. They contain small amounts of tanning substances, volatile oils, pectin substances, anthocyanin compounds, salts of iron, phosphorus, calcium, cobalt, manganese, and potassium.

Intensive technologies with the use of highly productive cultivars, dense planting schemes, and shortened plantation rotation are used in current strawberry cultivation. Production systems are very important for the post-harvest quality of the berries, including storage terms, susceptibility to gray mold, fruit firmness, and polyphenolic compound content [1].

It is a known fact [2] that the application of mulching while growing strawberries increases soil temperature by 0.5–2.0°C at the depth of 5–10 cm, enhances berry ripening by 2–11 days, and facilitates the number of harvesting by 1–4. The use of black polyethylene as mulch helps increase plant biomass by 33 percent, yielding capacity increases by 16.4–26.1 percent, and marketable output increases by 9.0 percent. Using grain straw as mulch increases yielding capacity by 9.6 percent and marketable output by 4.3 percent [3]. The findings by Fan L. et al. [4] prove that plastic mulch with the covered rows improves strawberry quality considerably, and the research done in Poland [5] indicates that mulching results in the increase of the strawberries (class 'Ekstra'). Sharma et al. [6] and Wang et al. [7] observed the connection between the increased fruit output per plant, their average weight, dry substance content and berry firmness, and increased soil temperature and quality of the radiation under the cover.

The application of mulching in strawberry production awakens interest as to storage terms, mass losses, and content of chemical composition components. Wang S. Y. et al. [8] studied the effect of mulch types on chemical composition of the strawberry plants. High content of soluble carbohydrates was recorded in the berries grown with the use of black polyethylene compared with that of red one and grain hay. Similar results were received in the trials carried out in Columbia [9]: the use of black polyethylene has a positive effect on dry soluble substances and titratable acidity.

The experiments carried out by Fan L. et al. [1] state that the use of plastic mulch decreases weight losses, sap leak, and the development of gray mold very much when strawberries are stored, compared with those grown under a matted system. The shine of these berries remains within 3 days after the harvest when they are stored at room temperature.

The purpose of our research was to evaluate the effect of mulching types on the change of quality indicators of the strawberries during the storage.

17.2   MATERIALS AND METHODOLOGY

According to the methodical recommendation as to the storage of fruits, vegetables, and grapes [10], the experiments were carried out in the refrigerator of the Technology storage and processing of fruits and vegetables Chair at Uman National University of Horticulture in 2011–2013; the berries of such cultivars as 'Festyvalna romashka' (control), 'Ducat,' 'Honey,' grown on various types of soil management, without mulching (control), mulching of the rows with black polyethylene and black agro-cloth, were used in the experiment.

Strawberries were harvested at a marketable stage, and selected in accordance with GOST 6828-89 [11]. Chilled berries of the first marketable cultivar were put into perforated plastic boxes (0.5 kg), then packed into polyethylene bags (the thickness of a film is 50–60 μm) and hermetically sealed. Berries were stored at temperature $0 \pm 1°C$ and relative air humidity 90–95 percent during 11 days. Natural mass loss, marketable indicators, and variation of a berry chemical composition were determined in the course of the trial.

Marketable analysis of the strawberries after the storage was made in compliance with GOST 6828-89; natural mass loss was defined by weighing fixed samples.

The content of dry soluble substances was determined by refractometric method (GOST 28562−90), [12] that of sugars by ferrocyanid method (GOST 8756.13–87), [13] acidity by titrating alkali (GOST 25555.0–82), [14] ascorbic acid by iodine-metric method (GOST 24556-89 [15] Statistic analysis was done using StatSoft STA-TISTICA 6.1.478 Russian, Enterprise Single User (2007).

17.3   RESULTS AND DISCUSSION

As the result of the experiments indicate, it was found that strawberries accumulated 9.5–10.6 of dry soluble substances, 5.6–9.5 percent of sugars, 0.75–1.1 percent of organic acids, 55.1–99.8 mg/100 g of ascorbic acid; it depended on the cultivar and cultivation condition (Table 17.1).

High content of dry soluble substances was typical for the strawberries grown without mulching—10–10.6 percent; this indicator for the strawberries grown on the mulched soil (black film) was lower by 0.5–1.6 percent. The strawberries of 'Honey' cultivar had the largest amount of dry soluble substances—10–10.6 percent among all the cultivars studied, which depended on the mulching type.

The strawberries grown without mulching accumulated the largest amount of sugars—6.4–9.5 percent, whereas the strawberries grown on mulched soil in the rows (agro cloth) had less sugars by 0.8–3.9 percent. The strawberries of 'Honey' cultivar had the largest amount of sugars—6.5–9.5 percent, depending on the cultivation conditions (Table 17.1).

TABLE 17.1 Natural mass loss, marketable output, and main components of the chemical composition of the strawberries in the process of storage

Cultivar	Natural mass losses (%)	Marketable output (%)	Mass share (%)			Ascorbic acid content (mg/100 g)
			Dry soluble substances	Sugars	Organic acids (on citric acid basis)	
Festyvalna romashka without mulching (control)	1.6	80.2	10.0	6.7	0.75	99.8
			9.1*	5.1	0.74	58.9
Festyvalna romashka (mulching with agro cloth)	1.9	74.6	9.9	6.0	0.95	74.8
			9.0	5.9	0.75	56.3
Festyvalna romashka (mulching with back film)	2.1	70.5	9.5	6.4	0.97	79.2
			8.8	5.4	0.65	57.2
Ducat without mulching (control)	0.9	83.1	10.0	6.4	0.99	57.1
			9.5	6.0	0.69	46.6
Ducat (mulching with agro cloth)	1.5	76.1	9.7	5.6	1.01	55.1
			9.3	6.2	0.66	46.6
Ducat (mulching with back film)	2.2	73.8	9.8	6.7	1.06	56.1
			9.0	5.5	0.68	46.6
Honey without mulching (control)	1.1	73.8	10.6	9.5	0.93	77.4
			9.5	6.1	0.70	45.3
Honey (mulching with agro cloth)	1.8	65.5	10.4	8.6	0.85	84.2
			9.2	5.9	0.64	51.0
Honey (mulching with back film)	2.0	63.2	10.0	6.5	0.89	72.6
			9.4	6.4	0.70	48.1
LSD$_{0.05}$	0.13	0.69	0.21	0.22	0.03	0.60

*Note: **above** the line—the content of chemical composition components of the strawberries before storage; below the line—the same after storage.

The content of organic acids in strawberries ranged from 0.75 to 1.06 percent. Their higher content was recorded in the strawberries grown on soil mulch in the rows with black polyethylene—0.89–1.06 percent which corresponds to the data of Casi-erra–Posada F. et al. [9]. The strawberries grown without mulching had a lower indica-tor—0.14–0.31 percent. Our earlier tests prove [16] that in the process of the straw-berry ripening the organic acid content decreases gradually. Thus, the organic acid content did not decrease to an optimal level in the strawberries grown on a mulched soil and when they developed red coloring. 'Ducat' strawberries were the tartest ones (0.99–1.06%).

The strawberries grown without mulching had high vitamin value—57.1–99.8 mg/100 g, whereas the use of agro cloth resulted in its decrease by 2–27.1 percent. The highest content of ascorbic acid was recorded in the strawberries of 'Festyvalna romashka' cultivar (74.8–99.8 mg/100 g). It should be noted that the use of mulching facilitated faster ripening of the berries, but it had a negative effect on the accumula-tion of the main components of the chemical composition.

During the storage because of berry respiration and selective gas permeability of polyethylene, a modified gas atmosphere with higher carbon dioxide content and lower oxygen content is formed [17]. When strawberries were stored in this atmosphere, we observed natural mass loss—from 0.9 to 2.2 percent. The highest mass loss was typi-cal for the berries grown on row mulching with black polyethylene—2.0–2.2 percent, depending on the cultivar; the lowest one, from 0.9 to 1.6, for control. Strawberries of 'Festyvalna romashka' cultivar were characterized with higher mass loss compared with others—from 1.6 to 2.1 percent.

The loss of organic substances occurs together with mass loss in the process of storage. Within 11 days, the content of dry soluble substances in the berries decreased by 4.5–11.3 percent as compared with their content before the storage. The lowest losses of dry soluble substances were recorded for the strawberries grown without mulching—0.2–1.6 percent, the losses were higher by 1–3 percent in the berries grown with the use of mulching.

The highest losses of dry soluble substances, depending on mulching type, were observed in the berries of 'Honey' cultivar was 5.9–11.3 percent; the same indicator for 'Festyvalna romashka' cultivar was 7.3–9.4 percent. The lowest losses—4.5–8.2 percent—were recorded for 'Ducat' strawberry.

The use of mulching in the strawberry rows made it possible to reduce the losses of dry soluble substances from 9.4 to 7.3–8.8 percent ('Festyvalna romashka' cultivar). The strawberries of 'Ducat' and 'Honey' cultivars, grown without mulching, showed the highest content of dry soluble substances after the storage—9.4–9.5 percent.

Within 11 days of the storage sugar content in the berries decreased by 2.3–35.5 percent. Among the cultivars studied, strawberries of 'Honey' cultivar showed the high-est losses of sugars by 2.3–35.5 percent, those of 'Festyvalna romashka' cultivar were 2.5–25.3 percent. The lowest losses were recorded for 'Ducat' strawberries—5.2–18.1 percent. The use of mulching enhanced the stabilization of sugar content in the berries during the storage. Their losses were smaller by 4.0–32.2 percent depending on the mulching type. The strawberries of 'Ducat' cultivar were characterized by high sugar content after the storage.

In the process of the storage, the decrease of organic acid content in the strawber-ries by 1.3–35.8 percent was observed. 'Ducat' strawberries showed the highest losses of organic acids—30.3–35.8 percent. Mulching in the rows with black polyethylene and agro cloth resulted in the increase of organic acid losses during the storage by 21.3–35.8 percent as compared with control.

The ascorbic acid content was decreased by 16.7–42.1 percent during the storage. The trials proved that the use of mulching during strawberry cultivation enhanced ascorbic acid losses during the storage by 9.2–20.5 percent. The highest losses of ascorbic acid were observed for the strawberries of 'Honey' cultivar—35.1–42.12 percent; they were lower by 0.2–15.9 percent for 'Festyvalna romashka' cultivar. The strawberries of 'Ducat' cultivar showed high ascorbic acid content after the storage—45.6–46.2 mg/100 g.

Marketable analysis of the berries after the storage showed that standard output depended considerably on soil management type when they were grown. The use of mulching while growing strawberries reduced the output by 5.6–10.6 percent as compared with control, the level of technical reject being higher by 3.9–5.4 percent.

17.4   CONCLUSION

The use of mulching while growing strawberries enhances the accumulation of smaller amount of dry soluble substances, sugars, and ascorbic acid, but berries have higher acidity. The strawberries, grown with soil mulching, had an increased loss of mass and organic substances and lower output of marketable produce during the storage. The lowest losses of mass and organic substances during the storage and high marketable output were recorded for the strawberries of 'Ducat' cultivar.

KEYWORDS

- **Anthocyanin**
- **Cultivar**
- **Storage**
- **Strawberry**
- **Vitamin**

REFERENCES

1. Fan, L.; Yu, C.; Fang, C.; et al., The effect of three production systems on the postharvest quality and phytochemical composition of Orleans strawberry. *Can J. Plant Sci.* **2011**, *91(2)*, 403–409.
2. Loginova, S. F.; Effect of Soil Mulching Dark Film on the Yield and Quality of Berries Strawberry Varieties: Author. Dis ... Candidate. Agricultural Sciences: 06.01.07. St. Petersburg: State. Agrarian. Univ. St. Petersburg; **2003**, 21 p. (in Russian).
3. Butsyk, R. N.; The Productivity of Strawberries Depending on the Plantations Covering, Mulching of Soil and Fertilizing in the Right–Bank Forest–Steppe of Ukraine: Author. Dis ... Candidate. Agricultural Sciences: 06.01.07. Uman: Uman National University of Horticulture; **2011**, 20 p. (in Ukrainian).
4. Fan, L.; Roux, V.; Dubé, C.; Charlebois, D.; Tao, S.; and Khanizadeh, S.; Effect of Mulching Systems on Fruit Quality and Phytochemical Composition of Newly Developed Strawberry Lines. Agricultural and Food Science; **2012**, *21(2)*, 132–140.

5. Frac, M.; Michalski, P.; and Sas–Paszt, L.; The effect of mulch and mycorrhiza on fruit yield and size of three strawberry cultivars. *J. Fruit Ornamental Plant Res.* **2009**, *17(2)*, 85−93.

6. Sharma, R. R.; Singh, R.; Singh, D.; Gupta, R. K.; Influence of row covers and mulching interaction on leaf physiology, fruit yield and albinism incidence in 'Sweet Charlie' strawberry (Fragaria x ananassa Duch.). *Fruits.* **2008**, *63(2)*, 103−110.

7. Wang, S. Y.; Zheng, W.; and Galletta, G. J.; Cultural system affects fruit quality and antioxidant capacity in strawberries. *J. Agric. Food Chem.* **2002**, *50(22)*, 6534−6542.

8. Wang, S. Y.; Galletta, G. J.; Camp, M. J.; and Kasperbauer, M. J.; Mulch types affect fruit quality and composition of two strawberry genotypes. *Hort Sci.* **1998**, *33(4)*, 636−640.

9. Casierra–Posada, F.; Fonseca, E.; and Vaughan, G.; Fruit quality in strawberry (*Fragaria* sp.) grown on colored plastic mulch. *Agronomia Colombiana.* **2011**, *29(3)*, 407−413 p.

10. Guidelines for the Storage of Fruits, Vegetables and Grapes. Organization and Research General Ed. Dzheneeva, S. Y.; and Ivanchenko, V. I.;. Yalta: Institute of Vine and Wine "Magarach," **1998**, 152 p. (in Ukrainian).

11. Requirements for State Purchases, Deliveries and Retail (GOST 6828-890) Fresh Strawberry. [Introduced from 01.01.1991]. Moscow; **1989**, 25 p. (in Russian).

12. Refractometric Method for Determination of Soluble Dry Substances Content (GOST 28562−90) Fruit and Vegetable Products. [Introduced from 01.07.1991]. Moscow; **2005**, 12 p. (in Russian).

13. Methods for Determination of Sugars (DSTU 4954:2008) Fruit and Vegetable Products [Introduced from 26.03. 2008]. Kyiv: Derzhspozhyvstandart; **2009**, 21 p. (in Ukrainian).

14. Methods for Determination of Titratable Acidity (DSTU 4957:2008) Fruit and Vegetable Products. [Introduced from 01.07.2009]. Kyiv: Derzhspozhyvstandart; **2009**, 10 p. (in Ukrainian).

15. Methods for Determination of Vitamin C (GOST 24556-89) Products of Fruits and Vegetables Processing. [Introduced from 1990-01-01]. Moscow; **2003**, 10 p. (in Russian).

16. Zamorska, I. L.; Dynamics of Accumulation of Biochemical Components of Strawberry Pineapple in the Right−Bank Forest-Steppe of Ukraine. Proceedings of Uman State Agrarian University; **2002**, *55*, 129−133 (in Ukrainian).

17. Zamorska, I. L.; Store Strawberries in Plastic Bags. Proceedings of Uman State Agrarian University. Part 1. Agriculture; **2008**, *67*, 123−127. (in Ukrainian).

CHAPTER 18

THE QUALITY OF SUGAR BEET SEEDS AND THE WAYS OF ITS INCREASE

VOLODYMYR AR. DORONIN, YAROSLAV V. BYELYK, VALENTIN V. POLISHCHUK, and LESYA M. KARPUK

CONTENTS

18.1 INTRODUCTION

To ensure the competitiveness of the sugar beet in the global market, it is necessary to create the conditions to ensure high technological level of growth, which meets the requirements of sustainable development, to implement high environmental standards for the sugar beet processing and to orient the consumers to their market. Many factors determine sugar beet productivity in the intensive agriculture: soil and climatic conditions, the implementation of high-productivity hybrids, the qualitative preseeding processing of seeds, the use of modern techniques and technologies, fertilizers, reliable plant protection, high-tech improvements on the factories, etc. All these factors can reduce the sugar beet productivity significantly, but it is not possible to achieve the maximum yield of culture without the use of high-quality seeds of new hybrids.

The quality of sugar beet seeds is determined by the complex of genetic factors that are controlled by plant breeders and agroecological and agrotechnical conditions of their growing and methods of post-harvest and preseeding seed preparation with modern technology use [1]. Therefore, we have focused only on those methods influencing directly the yield and quality of sugar beet under its cultivation with and without plantings methods and in the process of its preseeding preparation. The most important indicators of seed quality are its viability, germination energy, field germination, one sprouting, uniformity, and stability in the size and forms.

Among the many factors influencing the growth, development, and yield formation and quality of seeds of great importance are the processes of controlled regulation of flowering and pollination of seed plants, especially under growing hybrids of seeds based on cytoplasmic male sterility (CMS). It is next to impossible to avoid the formation of a large number of small seeds that under the current standard [2] do not refer to the seeds and is lost by the post-harvest purification of heap without solving the problem of the methods improvement of the directional regulation of growth processes. Also, these methods aim at limiting the growth of tall plants, which improves the conditions of seed harvesting, reduces the losses, and increases the yield and quality.

To limit growth of sugar beet seed plants, they use manual, mechanical, and chemical minting. This method provides more productive seed plants through central stem growth limiting, which results in the more active entry of side stems nutrients, which improves their growth and development and, ultimately, increases their productivity [3].

With the removal of an apical meristem of the central stem, its growth and development is suspended. But not only the point of growth under the minting is removed, some part of the stem with fruit on it is removed as well; as a result, redistribution of nutrients and other substances necessary for the growth of both central and side stems takes place. Instead of entering the point of growth for central stem growth, development, and new small fruits formation, these substances enter the fruit remaining on the stems of seed plants. Application of minting forms larger seeds, and it accumulates more nutrients as well. Application of this agriculture method in the early phase of stem formation accelerates the start of seed plants flowering within 2–3 days. This method is easier and ends at earlier term, which in turn accelerates the maturation of seeds within 2–3 days. Also, a positive effect of the minting on the seed quality, especially when the minting is done in the late phase of stem formation, is observed [4].

According to V. Faydyuk, the yield of Ukrains'kyi ChS 70 hybrid seed is increased by 0.09 t/ha, similar to 3 percent under applying seed plants minting [5]. Odessa region, by both manual and mechanized minting in the early phase of stem formation, has shown an increase in the yield of seed by 0.15–0.35 t/ha, compared to the control—without minting under planting way of seed growing [6]. In irrigated conditions of Crimea at "Zarichniy farm" on an area of 1.1 ha, the minting in the phase of mass stem formation has provided the increase of the seed yield of varieties of populations by 0.57 t/ha, and similarly by 3 percent [7]. Thus, the previous researches conducted on growing ordinary seeds of the varieties-populations and sugar beet hybrids based on CMS both with plantings and without plantings ways demonstrate the high efficiency of minting as the method of directed regulation of seed plants growth and development and their flowering, pollination, and fertilization. The modern hybrids of sugar beet have high potential of seed productivity, but for its more complete implementation they should create favorable conditions for the growth of mating components. Therefore, studying the growth process of mating components and development regulation, synchronization of its flowering and formation of base seed yield by the criterion of maximum seed productivity was topical.

One of the most perspective ways of improving the seed quality is its pre-sowing preparation in the seed plants, which includes seed cleaning from impurities that do not relate to the main crop seeds, sizing, polishing, sorting by the aerodynamic properties and specific gravity, stimulation, pelleting, and encrusting. The stimulation of the intensity of seed germination is possible with the use of mechanical methods of seed preparation on the seed plants by way of removing the artificial barriers to seed germination, the use of growth stimulants, and microelements. However, the most perspective way to increase the intensity is to initiate the passing of germination start phase with its follow suspension, which was the goal of our research.

18.2 MATERIALS AND METHODODOLOGY

The research program envisaged studying the specifics of seed quality formation both under seeds growing and its pre-sowing preparation in the seed plants. The research was conducted at the Institute of Bioenergy Crops and Sugar beet of NAAS, Uman Experimental Breeding Station, Uman National University of Horticulture, and Vinnytsa seed factory Ahrohrad V Ltd in 2008–2013.

The field experiments were conducted according to the following scheme:

1. No minting—control;
2. 50 percent minting of plant pollinator;
3. 50 percent minting of plant pollinator and 100 percent of plant CMS component.

The study of the optimal terms of minting and its efficiency in the processes of flower formation, the flowering synchrony, and seed plant productivity was performed simultaneously on the paternal and maternal components Umans'kyi ChS 97 sugar beet triploid hybrids. The minting was carried out manually in the period of mass stem formation when the plants were 60–70 cm in height. At the same time, we removed the top of the main stem for 5–10 cm. The area of scoring plot was 56 m², with triple repetition.

To stimulate the seed under the production conditions, 12 parties of calibrated seed of diploid hybrids Ukrains'kyi ChS 72, Vesto and triploid Dobroslava, Oleksandriia were used. The stimulation of seed was performed by the method of Institute of Bioenergy Crops and Sugar Beets. Nonstimulated seed was sown in the control variant.

We determined in the laboratory conditions germination energy, field germination, and seeds purity [8], the mass of 1,000 fruits and one sprouting and one seeding [9]. The selection of the average seed samples was performed in accordance with applicable State standards of Ukraine (DSTU) [10]. The number of flowers by the experiment variants was determined by calculation; the seeds yield was determined by weighing the heap from the calculation plot and from the individual seed plants in the field conditions on the seed plants. The statistical processing of the experimental data was carried out by the methods of R.A. Fisher [11] with the appropriate computer software.

18.3   RESULTS AND DISCUSSIONS

The research has proved that the minting of seed plant-mating components has a positive effect on the processes of growth and development and on the flowering synchronicity and flower formation in particular. The suspension of the stem tops growth results in redistribution of nutrients, improves the supply of flowers formed, which contributes to additional formation of high-quality seeds and thus an increase in its productivity.

The experiments are conducted with the plants of CMS component and Umans'kyi ChS 97 triploid hybrid O-type in tunnel isolation (Figure 18.1).

FIGURE 18.1   The flowering of basic components in the tunnel isolation (after the minting in the depths of picture and without it in the foreground).

Minting is free (control) in the variant; the variability of the number of flowers, during the flowering, by the date of accounting was in the CMS component from 58.5 to 806.7 pc./plant in the sterility fixator (O-type) from 124.5 to 939.9 pc./plant (Figure 18.2(a)).

During all the dates of account of O-type, there were more flowers than in the MS component, which indicates no synchronicity of flowering of mating components, and which ultimately affects negatively the degree of tying seeds, its germination, and seed production.

The minting of 50 percent of the plants of O-type somewhat reduced the intensity of its flower formation during the whole period of flowering, compared to the control (Figure 18.2(b)). Thus, at the beginning of flowering 124.5 flowers/plants were formed without minting and 98.4 with minting. Similar results were obtained in the last date of accounting.

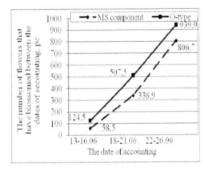
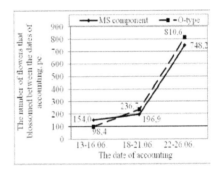

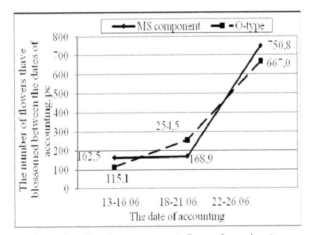

FIGURE 18.2   The intensity of mating components flower formation (average of 2008–2010) (a) without—control (b) under 50 percent of the plants of O-type (c) under 50 percent of plants of O-type and 100 percent of plants of CMS component.

There is a significant increase in the intensity of flower formation in the early phase of the stem formation and a slight decrease in the other two phases—154.0–748.2 pc./plant in the plants of CMS component without their minting. That is, the minting of 50 percent of plants of O-type ensured the synchronous flowering of both components at the beginning of flowering, and at its end.

In the variant of 50 percent of the plants of O-type minting and 100 percent plants of its sterile analogue (MS component), there was a negligible deviation of variation of the number of flowers of the two components that makes 115.1–667.0 pc./plant for a O-type and 162.5–750.8 pc./plant for the CMS component (Figure 18.2(c)).

Thus, minting of both plants of the O-type and of both components of breeding has provided the synchrony of flower formation and their flowering. Without the minting, the flowering of seed plants of O-type began and was finished in 2.1 times earlier, and it took place more intensively at the beginning of flowering and 1.3 times at the end of flowering, than that of the CMS component seed plants, that is, the flowering of mating components was not synchronous (Figure 18.3(a)).

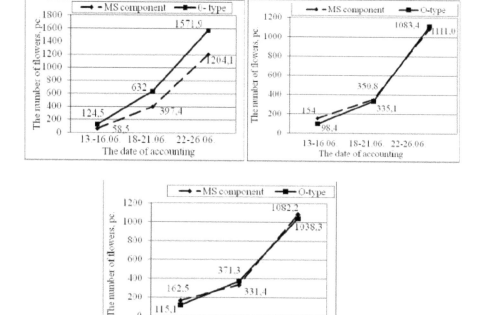

FIGURE 18.3   The dynamics of mating components flowering (average of 2008–2010) (a) without—control, (b) under 50 percent of the plants of O-type, and (c) under 50 percent of plants of O-type and 100 percent of plants of CMS component.

The minting of 50 percent of the plants of O-type has ensured the extension of its flowering term and more synchronous flowering of the components (Figure 18.3(b)). The flowering of O-type was only in 0.64–1.03 times more intensively. Both at the beginning of flowering and its completion, the number of flowers of CMS component and O-type were almost the same. Thus, by the last date of accounting, the number of flowers that bloomed in the CMS component were 1082.2 pc/plant, in O-type—1038.3 pc./plant.

It was established in the process of studying the influence of plant minting of diploid CMS component on seed production that this method contributes to a significant increase in the yield of the baseline seed (Table 18.1).

TABLE 18.1 The yield of baseline seed of CMS component increased to 0.17 t/ha under minting 50% of the plants of O-type.

Variant	Yield of seed (t/ha)	The degree of tying (%)	Energy of germination (%)	Field germination (%)	Mass of 1,000 pc (g)
Without minting— control	1.47	86.6	76	81	12.0
The minting of 50% of pollinator	1.64	91.0	85	87	13.1
The minting of 50% of pollinator and 100% of CMS component	1.67	91.0	87	90	13.1
LSD_{05}	0.11	1.4	3.3	3.5	0.7

By minting 50 percent of the plants of O-type and 100 percent of plants of CMS component, the yield increased by 0.20 t/ha compared to the control. That is, both the methods of minting provide significant increase of CMS component seeds yield, but there is no significant difference between them.

An important factor influencing the seed productivity indexes and, especially, the seed quality is its degree of tying, which depends on the synchronicity of the components of hybrid flowering. In our research, this feature is varied in the range from 86.6 to 91.0 percent. Thus, the degree of seeds tying was affected with the minting as an O-type and both parental components as well. Providing the synchronization of components flowering contributed more to substantially increase the degree of tying seeds compared with the control that influenced on its germination. Thus, the minting of only 50 percent of the plants of O-type ensured 6 percent increase in seed germination compared with the control, and the minting of both components of 50 percent of the plants of monogerm seed O-type and 100 percent of plant of CMS component provided receiving of higher seed germination—90 percent with the control variant index of 81 percent. In addition to the increase of yield and seed germination, the mass of 1,000 fruits increased significantly, which indicates actively entering the nutrients forming the side stems.

Similar results were obtained with the yield and seed quality of O-type that depend on the directional regulation of flowering process of the hybrid components. The yield of seed increased to 0.17–0.20 t/ha and seed germination increased from 82 percent (control) to 87–90 percent (in variants with minting). The highest index of germination is observed in the variant, where the minting was applied on 50 percent of plants and 100 percent of plants of CMS component (Table 18.2).

TABLE 18.2   The yield and quality of O-type basic seed depending on the directional regulation of the process of flowering depending on seed plant minting (average of 2008–2010)

Variant	Yield of seed (t/ha)	The degree of tying (%)	Energy of germination (%)	Germi-nation (%)	Weight of 1,000 pc (g)
Without mint-ing—control	1.48	88.8	78	82	12.2
The minting of 50% of pollinator	1.65	93.4	85	87	13.3
The minting of 50% of pollinator and 100% of CMS component	1.68	93.7	87	90	13.5
LSD$_{05}$	0.04	1.7	2.6	3.1	0.4

The increase of seed germination is caused by better pollination of O-type plants, as evidenced by the degree of seed tying, which increased by 4.6–4.9 percent compared with the control. The minting positively impacts on the mass of 1,000 seeds of CMS component and O-type. The significant increase of the mass of 1,000 fruits in variants with directional regulation of the process of flowering is caused by better redistribution of nutrients.

The restriction of the central stem growth results in the more active nutrients entering the side stems, where the basic mass of the seeds is formed, improving their growth and development and, ultimately, decreasing the number of fruits of the diameter less than 3.50 mm, which does not relate to the seeds and is lost in the post-harvest treatment of heap. It is found out that under regulation of the process of flowering the number of fruit with the 3.00–3.50 mm diameter fraction decreased by 1.7–2.0 times of CMS component and by 1.9–2.2 times of O-type and the yield of sown fraction of seeds increased accordingly—by 8–10.4 and 9.7–11 percent (Table 18.3). Thus, without minting the number of seeds 3.00–3.50 mm fraction of CMS component was 24.3 percent, while both under minting only of 50 percent of the plants, it was 14.3 percent, and under minting of 50 percent of plants of O-type and 100 percent of plants of CMS component was 12.3 percent.

TABLE 18.3 The influence of regulation of the flowering and fertilization process on the fractional structure of seeds depending on seed plant minting (average for 2008–2010)

Variant	The content of fractions of seed (mm, %)			
	> 5.50	4.50–5.50	3.50–4.50	3.00–3.50
CMS component				
Without minting—control	3.7	14.7	57.3	24.3
The minting of 50% of pollinator	5.7	17.7	62.3	14.3
The minting of 50% of pollinator and 100% of CMS component	5.3	18.7	63.7	12.3
O-type				
Without minting—control	4.3	12.3	60.7	22.7
The minting of 50% of pollinator	5.7	16.7	66.0	11.7
The minting of 50% of pollinator and 100% of CMS component	5.7	16.7	67.3	10.3

By minting both components, we obtained the highest yield of sown seed fractions—82.4 percent. Under minting of 50 percent of the O-type, we observed somewhat lower yield of sown seeds fraction compared with the minting of both components due to a high content of fine fraction—less than 3.50 mm. Similar results were obtained with the O-type.

Thus, the directed regulation of processes of flowering and flower formation of seed plant mating components has a positive effect on the processes of growth and development, particularly on the synchronicity of flower formation, flowering, and seed tying degree and consequently on its yield and quality. The yield of seed and its quality increased significantly compared with the control (without minting) as a CMS component and O-type. Along with the yield increase, the yield of seeds sown fractions increases due to reduction of fruits, with a diameter of less than 3.50 mm.

In the process of growing sugar beet seeds, one cannot achieve the desired results on the quality of seed under due to high degree of its quality diversity caused by the biological characteristics of the crop (the phase of sugar beet seed plants flowering takes place not evenly during 20–40 days depending on the weather conditions in cultivation areas). That is why all breeding and seed companies in the world and in our country as well prepare of the sugar beet seeds for sowing in the seed plants only. During the pre-sowing preparation, it is a very difficult technological chain including the seed stimulation. All technological operations aim at receiving the maximum of seed quality.

The program of scientific-research work of Institute of Bioenergy Crops and Sugar Beet aims on the establishing factors contributing to rapid seed germination at low temperatures and developing method of seed preparation on seed plants with high germination and reliability, which ensures high field germination and accelerated development of young plants in the field. It is known that the field germination of seeds depends on many factors and, primarily, on the laboratory germination of the sown seeds, which, in its turn, depends on several factors, biological features of a hybrid, soil and climatic conditions of seeds growing, and post-harvest and presowing preparation.

As we know, not all of the seeds give stairs after sowing. According to the data by Ovcharov K.E. [12], some species need shell removal for seeds to germinate, others—inhibitors reducing the content, some—metabolites enrichment, and for others the influence of water, light, temperature, and other physical factors is necessary. The response reaction of seeds on the above-mentioned actions depends on the natural species of seeds and their physiological state, and conditional germination. Almost all of the above-mentioned methods of increasing germination intensity are applied to sugar beet seeds.

We applied two methods of stimulating increase of the intensity of germ germination of sugar beet seeds, mechanical way—by reducing the mechanical obstacle—seed pericarp, which is achieved by seeds polishing and by the initial phases of germination with its following suspension passing initiation. The latter is one of the most perspective ways of increasing the intensity of seed germination.

The process of studying the efficiency of stimulation by the mechanical way in order to reduce the seed injury and to increase the degree of seed polishing was carried out in stages.

It was established that in the process of uncalibrated seeds polishing, the removal of 26.7 percent of the mass of pericarp ensures a significant increase in the intensity of seed germination (Table 18.4).

Thus, in 48 h after seeding, 32 percent of fruits sprouted, which is 19 percent more than in the control, where the seeds are not polished. The similar dependence is observed in 72 and 96 h after seeding. Bsides even in 120 h after sowing the difference in the number of sprouted seeds was significant.

The repeated sequential polishing of seeds ensures the removal of pericarp mass up to 30.1 percent compared with the control, which contributed to the increase of its germination intensity especially in the early stages. The removal of 31.7 percent of pericarp mass causes a slight injury of the seed that does not affect its germination reducing intensity.

TABLE 18.4 The intensity of seed germination depending on the degree of polishing (average of 2011–2012)

Variant	Pericarp removed (%)	Germinated seeds in hours after sowing (%)				
		48	72	96% energy of germination	120	Germination
Control, the original sample	0	13	60	76	77	78
After 1 polishing	26.7	32	80	82	82	83
After 2 polishing	30.1	56	86	87	87	88
After 3 polishing	31.7	52	87	88	88	89
LSD$_{05}$	2.5	5.6	2.8	3.4	3.2	3.3

The final stage of seed preparation for pelleting is seeds polishing by the technological fractions and its sorting by specific mass. It was established that under seed polishing by the technological fraction with diameter of less than 3.75 mm, 5.3 percent of pericarp mass is removed and the intensity of germination in 72 h after seeding increases by 17 percent compared with the control (Table 18.5).

TABLE 18.5 The intensity of calibrated seeds germination depending on the degree of polishing (average of 2011–2012)

Variant	Pericarp removed (%)	Germinated seeds in hours after sowing (%)				
		48	72	96% energy of germination	120	Germination
Control, the original sample	0	13	60	76	77	78
Fraction of seeds < 3.75 mm in diameter						
Before polishing	0	24	71	74	74	74
After polishing	5.3	50	77	78	79	79
After the pneumatic table	0	34	98	99	99	99
Fraction of seeds > 3.75 mm in diameter						
Before polishing	0	27	85	88	89	89
After polishing	8.8	57	91	91	91	92
After the pneumatic table	0	59	98	98	98	99
LSD$_{05}$		3.3	2.4	2.7	2.5	2.5

The sorting of this seed according to its specific mass ensures obtaining calibrated seed with 99 percent germination, which is quite suitable for the preparation of high-quality pelleted seed. The similar results were obtained after the polishing and sorting by specific mass on the seed of technological fraction diameter more than 3.75 mm.

Along with the mechanical method of the seed germination intensity increase, we studied the possibility to improve it through the stimulation of the initial phases of passage of germination with its subsequent suspension.

To determine the optimum regime of stimulation, the research was carried out with the seed of two domestic triploid hybrids. As the results prove the hybrids Oleksandriia and Umans'kyi ChS 97 reacted to the seed to the stimulation in different ways. Under the stimulating initial phases passage of germination of hybrid Oleksandriia, the optimal term at which the number of sprouted incrusted seeds through the 48 h after seeding was 22 percent higher, and its germination was 6 percent higher than on the control, which is the stimulation by the fourth regime. At the stimulation of the initial phases passage of germination of hybrid Umans'kyi ChS 97 under all regimes, we did not obtain a positive result on seed germination. The intensity of germination in 48 h after sowing increased to 6–44 percent compared with the control. The essence revealed in the chapter as the materials for patenting are being prepared and only the results of laboratory tests are shown.

It is established that pelleting shell together with the protective preparations provides the mechanical barrier for seed germination, especially in the early stages—48 h after sowing. Under these conditions, the stimulation of seeds before its pelleting affects the intensity of pelleted seeds germination positively. In all the variants with stimulation, except the variant where the stimulation was performed under the first regime, the number of germinated seeds of both hybrids was higher than in the control.

The analysis of factors influencing the number of sprouted encrusted pelleted seeds in 48 h has showed that the share of factor "stimulation" impact is significantly increased, compared with the impact on the calibrating seed and made 30 percent, a hybrid factor—60 percent, other factors—10 percent (Figure 18.4).

The verification of the developed method of seed germination stimulation, which provides the awakening of germ in the early stages with its subsequent suspension in a production environment has confirmed the results of laboratory research on the efficiency of this method. The effectiveness of stimulation was checked in 12 parties of calibrated seed of diploid and triploid hybrids prepared for pelleting, which lost partially the energy of germination and germination in the process of storage.

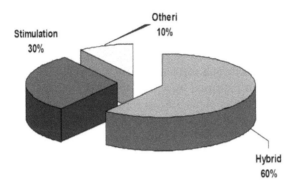

FIGURE 18.4 The share of factors influencing number of sprouted encrusted pelleted seeds in 48 h.

The seed stimulation ensures significant increase in the intensity of its germination of different sugar beet biological forms in the laboratory conditions (Figure 18.5.)

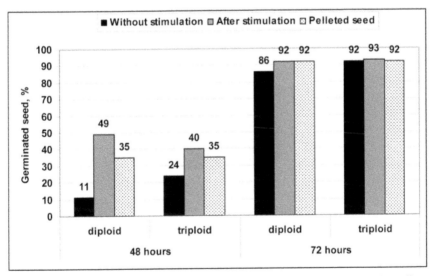

FIGURE 18.5 The intensity of seed germination of different biological forms depending on its stimulation (average of 12 parties of seeds, 2012).

Thus, if in 48 h after sowing 11 percent of calibrated seeds sprouted, then after the stimulation of diploid hybrids 49 or 38 percent more than in the control. Similar results were obtained by the triploid hybrids. After stimulated seeds pelleting, the intensity of its germination was significantly higher in the diploid and triploid sugar beets. Even 72 h after seeding, the intensity of calibrated and pelleted seed of diploid hybrids germination after the stimulation was higher than in the control. In triploid hybrids,

72 h after sowing, the difference in the number of germinated seeds was almost not observed or equals the control.

It is found that calibrated seed stimulation is affected significantly by the increase on its germination energy on both the biological forms of sugar beet. Thus, if the germination energy of calibrated seeds of diploid and triploid hybrids before stimulation was 90 percent, it increased by 4 percent (LSD$_{05}$ stimulating factor = 1.2%) after stimulation and made 94 percent. Seed germination significantly increased after stimulation of both biological forms of sugar beet. No significant differences in germination energy and germination seed was found out depending on seed parties studied as diploid and triploid hybrids. After stimulated calibrated seed pelleting, the germination energy and germination of both the biological forms of beets were equal to those before the pelleting, but significantly higher compared with control. Thus, the germination energy and germination of pelleted seeds of diploid hybrids were identical and made 95 percent, which is 5 and 3 percent high than in the control, respectively, and in triploid hybrids these indexes of pelleted seeds were equal and made 94 percent, which is 4 and 2 percent more than in the control.

It was found out while determining the factors influencing on the laboratory germination of seeds that the "stimulation of seed" factor was the most significant and made 61 percent (Figure 18.6).

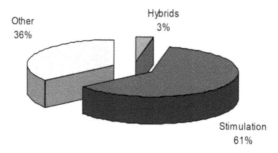

FIGURE 18.6 The share of factors influencing on seed germination (average of 12 parties of seeds, 2012).

The influence of biological forms of beet was insignificant and made only 3 percent, and the influence of other factors (presence of filled but dead fruit and others) was significant and made 36 percent.

18.4 CONCLUSIONS

-   The directed regulation of flowering processes and flower formation of seed plants of mating components has a positive effect on the processes of growth and development and, especially, on the synchronization of flowering and the degree of tying seed and consequently its yield and quality. The yield of seed and its quality are significantly increased compared with the control (without minting) as a CMS component of the O-type.

- The increase in seed yield is caused by the increase of 1,000 seeds mass and decreases the seed in diameter less than 3.50 mm. Under the minting, we observed a higher yield of seed and sown fractions 3.50–4.50; 4.50–5.50 mm, especially under the minting of both components—50 percent of the plants of O-type and 100 percent of plants of CMS component.
- To regulate the process of plant growth and development, the components of the crossbreeding is advisable to perform the minting of 50 percent of fixative plant and 100 percent of plants of CMS component, which provides the largest synchronicity of flower formation and flowering of hybrid components and the productivity of seed plants.
- It is found that in the process of uncalibrated seed polishing, the removing of 24.7 percent of mass pericarp has provided significant increase in the intensity of seed germination. In 48 h after sowing, the number of seeds sprouted increased by 32 percent compared with the control, where the seeds are not polished.
- It is proved that the pelleting shell together with protective agents create a mechanical barrier to seed germination especially in the early stages—in 48 h after sowing. Under these conditions, the stimulation of seed before pelleting effects is positive on the intensity of the pelleted seed germination. In all the variants with the stimulation, except the variant where the stimulation was performed within 2 h with the moisture content of the seed 35 percent, the number of germinated seeds of both hybrids was higher than in the control.
- The optimum time of stimulation of the initial phases passage of germination of hybrid Oleksandriia under which the number of sprouted encrusted tablets in 48 h after sowing was 22 percent higher and the germination energy and seed germination, respectively—by 7 and 6 percent higher than in the control was the fourth regime of stimulation.
- After the stimulated calibrated pelleting seed, the germination energy and germination of both the biological forms of beet was equal to those before pelleting, but significantly higher compared with the control.

KEYWORDS

- **Seeds**
- **Sugar beets**
- **Minting**
- **Additional pollination**
- **Seed stimulation**
- **Germination**
- **Yield**

REFERENCES

1 Doronin, V. A.; Biological Basis of Sugar Beet Hybrid Seed Formation and Ways Improve its Yield and Quality: Dissertation of Doctor of Agricultural Sciences: 06.01.14 Kyiv; **2003**, 305 p. (in Ukrainian).

2. Royik, M.V.; Balan, V.M.; Doronin, V.A.; et al.; Requirements for the Storing Sugar Beet Seed: DSTU 4231-2003. [Introduced from 01.01.2004] Kyiv: State Consumer Standard of Ukraine; **2004**, 5 p. (National Standard of Ukraine). (in Ukrainian).

3. Balan, V. M.; Tarabrin, A. E.; and Korneichuk, A. V.; No Drop-Off Testes Root Crops Biology and Agricultural Machinery in the Irrigated Conditions of the South of Ukraine. Ed. Balan, V. N.; Kyiv: Nora-Print; **2001**, 350 p. (in Russian).

4. Yukhnovsky, I. A.; Biological Features and Productivity CMS Hybrids Components Crossing Depending the Conditions of Cultivation. Vector-Agro: The Collection of Proceedings of the Institute of Sugar Beets; **2003**, *5,* 128–132 p. (in Ukrainian).

5. Faydyuk, V. V.; Hybrid seeds yield and quality depending the technology of its cultivation. Vector-Agro: The collection of proceedings of the Institute of Sugar Beets; **2003**, *5,* 134–135 p. (in Ukrainian).

6. Zarishnyak, A. S. and Levchenko, A. G.; The no drop-off testes growth limiting and their productivity. Sugar Beet; **1996**, *10,* 15–18 p. (in Russian).

7. Balan, V. M.; Salohub, Y. M.; Faydyuk, V. V.; and Yukhnovsky, O. I.; Hybrid Seeds Formation Under the Different Growing Conditions. Sugar Beets; **2003**, *3,* 8–9 p. (in Ukrainian).

8. Musiyenko, A.A.; Kobko, O.V.; Kuznechikova, V.M.; and Bidulia, K.H. Methods of Determination of Seed Germination, Monogermity and Quality. Sugar Beet Seeds: DSTU 2292-96 (GOST 22617.2-94) [Introduced from 01.01.1996] Kyiv: State Consumer Standard of Ukraine; 1995, 8 p. (National Standard of Ukraine). (in Ukrainian).

9. Royik, M.V.; Balan, V.M.; Doronin, V.A.; et al. Methods of Definition of the Mass of 1000 Seeds and Mass of one sowing Unit Requirements for Storing. Beet Seeds: [Introduced from 01.10.2004] Kyiv: State Consumer Standard of Ukraine; 2004, 15 p. (National Standard of Ukraine). (in Ukrainian).

10. Royik, M.V.; Balan, V.M.; Doronin, V.A.; et al. Rules of Acceptance and Methods of Specimen Selection. Sugar Beet SeedsDSTU 4328-2004. [Introduced from 01.07.2005] Kyiv: State Consumer Standard of Ukraine; 2005, 8 p. (National Standard of Ukraine). (in Ukrainian).

11. Fisher, R. A.; Statistical Methods for Research Workers. New Delhi: Cosmo Publications; **2006**, 354 p.

12. Ovcharov, K. E.; Seed Germination Physiological Basis. Moscow: "Nauka" (Science); **1969**, 280 p. (in Russian).

CHAPTER 19

OF COMBINING ABILITIES OF MALE STERILITY LINES AND STERILITY BINDERS OF SUGAR BEETS AS TO SUGAR CONTENT

MAKSYM M. NENKA and MYROSLAVA O. KORNEEVA

CONTENTS

19.1 INTRODUCTION

Sugar content is an important element of the productivity of sugar beet, which is a selection goal in the development of hybrids based on cytoplasm male sterility. Many scientists pointed out that sugar content is characterized by a significant variation factor (from 15 to 21%), which was significantly lower compared to yield [1, 2]. Studying the variability of populations of different origins, we found that the populations of the same variability were characterized by different absolute values of the sugar content, and vice versa [3]. Some scientists pointed at the appearance of transgress forms in the offspring with a frequency 0.7–1.4 percent [4].

Variability of sugar content depends on either the genotypic factors or the conditions of the environment and their interaction. The variability of this feature in populations depends mainly on the additive gene effects; in interline hybrids—on the additive and nonadditive effects [5–7]. However, the phenotypic expression of the sugar content is influenced a lot by other factors (environmental, agronomic, and others) that "mask" genetic parameters contributing to this feature, and create difficulties in the selection of genotypes.

"Cell method" (hexagonal method of organizing plants) with the intensity of selection of 15 percent was used to equalize differences caused by the environment in selection of some crops. Many authors indicated modifications in the areas of supply of sugar beet. Hence, A.L. Mazlumov thought that the use of the extended area of supply could identify all capabilities of the genotypes secured by nature [8]. He wrote that the extended area of supply influenced the variability of useful traits of beets more than special properties of soil or fertilizers. According to other researchers, it was shown that sugar content was higher in the offspring selected in the extended area of supply of the pioneers than in the selection on the normal area of supply. Moreover, the expansion of the phenotypic variance into components—genotypic and environmental—showed that the proportion of the genotypic variance to total phenotypic variance was higher in the extended area of supply [9].

19.2 MATERIALS AND METHODS

Maternal components are represented by two types—simple sterile hybrids (SSH) derived from crosses of sterile (MS) lines with unrelated binder of sterility (BS) and MS line—analogues of O type—was tested in the experiments in 2011–2012, conducted at the Institute of Bioenergy Crops and Sugar Beet in different environments. Backgrounds were as follows: normal background of fertilizing—the common area of supply (NBCA), normal background—the extended area of supply (NBEA), increased background of fertilizing—the common area (IBCA), and increased background—the extended area of supply (IBEA).

19.3 RESULTS AND DISCUSSIONS

1. ***Sugar content of SSH in the environments NBCA and NBEA.*** As the analysis of sugar content revealed (Table 19.1), SSH was characterized by specific reaction to changes in the area of supply in the environments NBCA and NBEA.

TABLE 19.1 Sugar content of SSH, deviation from average and standard environments of NBCA and NBEA, IBCSB of NAAS, 2011–2012

Simple sterile hybrids	Area of supply					
	NBCA			NBEA		
	Sugar content (%)	Deviation from the average (%)	Deviation from St (%)	Sugar content (%)	Deviation from the average (%)	Deviation from St (%)
MS 1/Ot 2	16.5	−0.5*	−4.3	17.8	0.8*	3.3
MS 1/Ot 3	16.7	−0.3	−3.3	16.0	−0.9*	−7.2
MS 1/Ot 4	16.6	−0.4*	−3.9	17.2	0.2	−0.2
MS 1/Ot 5	16.5	−0.4*	−4.1	18.1	1.1*	4.8
MS 2/Ot 1	17.3	−0.4*	0.5	17.2	0.2	−0.2
MS 2/Ot 3	16.9	−0.1	−2.2	15.9	−1.1*	−7.7
MS 2/Ot 4	17.1	0.1	−0.8	16.2	−0.8*	−6.2
MS 2/Ot 5	17.6	0.6*	1.4	17.5	0.6*	1.7
MS 3/Ot 1	17.3	0.4*	0.5	17.8	0.8*	3.3
MS 3/Ot 2	16.5	−0.5*	−4.5	16.3	−0.6*	−5.2
MS 3/Ot 4	17.3	0.4*	0.5	16.9	−0.1	−2.1
MS 3/Ot 5	17.8	0.8*	3.0	17.2	0.2	−0.2
MS 4/Ot 1	16.9	−0.1	−1.8	17.4	0.4*	0.8
MS 4/Ot 2	16.6	−0.4*	−3.9	16.5	−0.5*	−4.3
MS 4/Ot 3	16.6	−0.4*	−3.7	16.2	−0.8*	−5.8
MS 4/Ot 5	17.5	0.5*	1.3	17.2	0.2	−0.2
MS 5/Ot1	17.1	0.2	−0.6	16.7	−0.3	−3.1
MS 5/Ot 2	16.4	0.6*	−5.1	17.7	0.7*	2.5
MS 5/Ot 3	17.1	0.2	−0.6	17.4	0.4*	1.0
MS 5/Ot 4	16.9	−0.1	−2.0	16.2	−0.7*	−5.8

Note: *—Substantially 5% of the significance level.

As summarized in Table 19.1, on NBCA, significant positive deviation from the average medium value was observed in five SSH, on NBEA—in seven, that is, extended area (EA) of supply as the factor contributed to the appearance of increased sugar content. There is a specific reaction of genotypes: some hybrids reduced the feature value on the extended area (MS 1/Ot 3, MS 2/Ot 3, MS 2/Ot 4, MS 3/Ot 4, MS 3/Ot 5, MS 5/Ot 1), and other hybrids were not sensitive to such factor as the area of supply, and some of them—MS 5/Ot 2, MS 1/Ot 2, MS 1/Ot 4, MS 1/Ot 5, MS 4/Ot 1, MS 5/Ot 2—increased sugar content. It agrees with the observations of A.L. Mazlumov [8], who wrote that the sugar content on different areas of supply varied in different ways. There are plants in which the sugar content in the extended area of supply does not change, and increases. Selection of lines with high sugar content on this background improved the material on this feature significantly.

EA supply contributed to the manifestation of high sugar content in the hybrid MS 1/Ot 5, which showed the highest value of the index (18.1%). The similar tendency was observed in the combinations with high sugar content MS 3/Ot 1 (17.8%) and MS 5/Ot 2 (17.7%), which did not show themselves in the control variant NBCA. The best hybrids MS 1/Ot 2, MS 1/Ot 4, MS 3/Ot 1, MS 5/Ot 2 increased the group standard for 2.5...4.8 percent in the sugar content.

The range of variation in the sugar content on EA was higher than on CA. The amplitude values of this index varied from 15.9 to 18.1 percent (EA) and 16.4...17.6 percent (CA), with the difference of 2.2 and 1.2 percent, respectively.

Phenotypic variability of hybrid sugar content was divided on the genotypic and environmental components by means of dispersive analysis. It turned out that the effect of the genotype in the variant on NBEA was greater than on NBCA, and was 86.6 versus 69.2 percent, respectively. It indicated a good differentiating ability of such factors as the extended area of supply for the manifestation of the genotype. Genotypic variance was also divided into components (Figures 19.1 and 19.2).

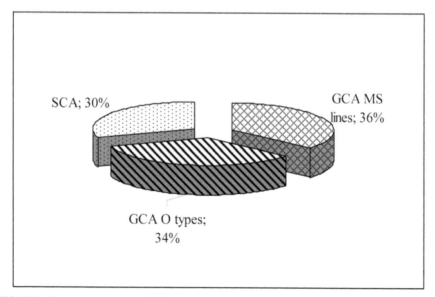

FIGURE 19.1   Genotypic variability and its share of sugar content feature of simple sterile hybrids, 2011–2012, environment NBCA.

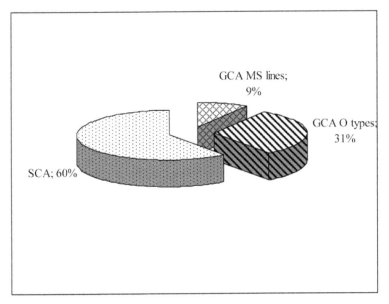

FIGURE 19.2   Genotypic variability and its share of sugar content feature of simple sterile hybrids, 2011–2012, environment NBEA.

A noteworthy fact is that the additive effects of genes of both parental forms (SCA of MS lines + SCA O types) on the extended area of supply were lower than on the common one and made the total of 40 versus 70 percent. While the share of nonadditive actions of gene was higher (60 vs. 30%). It shows that effects of component interactions became more significant on the extended area of supply (Figures 19.1 and 19.2).

Significant influence of parental forms and their interactions in the environment of NBCA revealed the effects of the combining ability—general (GCA) and specific (SCA) (Table 19.2).

TABLE 19.2   The effects of GCS and SCA of MS lines and O types, NBCA, 2011–2012

MS lines	Effects of GCS MS lines	Effects of SCA				
		Ot 1	Ot 2	Ot 3	Ot 4	Ot 5
MS 1	−0.38*	0	−0.16	0.46*	0	−0.31
MS 2	0.24*	0.05	0	0.04	−0.09	0
MS 3	0.27*	0.01	−0.39*	0	0.11	0.27
MS 4	−0.06	−0.05	0.04	−0.29	0	0.30
MS 5	−0.07	0.15	−0.15	0.25	−0.26	0

Note: *Substantially 5% of the significance level.

MS 2 and MS 3 lines were carriers of additive genes in the environment of NBCA. Significantly high effect of SCA was observed in the combination of MS 1/Ot 3 (+0.46*), but hybrid combination with their participation did not show competitive heterosis (deviation from St was—3.3%) because of the low effect of GCS of MS lines (−0.38*) (Table 19.1).

Combination of MS 2/Ot 5 (17.6%) had a significant difference of sugar content from the average population values (Table 19.1), which is due to the additive effect of the mother form—MS 2 (+0.24*) (Table 19.2).

Dispersive analysis of the data showed that the effect of all components of the variation of genotypes was significant in the environment NBEA. This allowed determining the proportion (Figure 19.2), as well as the effects of combining abilities of parental lines of SSH (Table 19.3).

TABLE 19.3 The effects of GCA and SCA of the sugar content of MS lines and O types, environment NBEA

MS lines	Effects of GCS MS lines	Effects of SCA				
		Ot 1	Ot 2	Ot 3	Ot 4	Ot 5
MS 1	0.30*	#	0.13	−0.78*	0.13	0.52*
MS 2	−0.27*	0.09	#	−0.31	−0.34	0.55*
MS 3	0.08	0.34	−0.23	#	0.01	−0.13
MS 4	−0.14	0.14	0.16	−0.40*	#	0.10
MS 5	0.03	−0.71*	1.15*	0.59*	−1.05*	#

Note: *Substantially 5% of the significance level.

Line MS 1 was the best among SCA in the environment of NBEA, which was well combined with MS Ot 5 (SCA = 0.52 *). Hybrid with their application had the highest sugar content—18.1 percent (Table 19.1). The components of hybrids MS 2/Ot 5, MS 5/Ot 2, and MS 5/Ot 3 had interaction effects that were significantly higher—0.55, 1.15, and 0.59, respectively, resulting in the increased level of sugar content in the hybrids, created with their participation—17.5, 17.7, and 17.4 percent, respectively.

Thus, nonadditive variance was dominated by EA compared with the common area (OA), and was 60 versus 30 percent on the common background of fertilizing in the genotypic structure of variability of the feature of sugar content.

Extended area of supply is the factor in which the effect of the genotype is higher than normal (86.6 vs. 69.2%), which indicates the feasibility of selecting the best genotypes in this environment. Differentiating ability of the environment NBEA is higher than on NBCA (seven of the best hybrids vs. five were distinguished). The range of variation in the sugar content on EA was higher (2.2%) compared to the CA (1.2%). Variability of effects of GSA and SCA, as well as the specificity of the reaction of hybrids due to the change of the area of supply, was distinguished. The best hybrids in the environment NBCA were hybrids MS 3/Ot 5 and MS 2/Ot 5, and in the environment NBEA—MS 1/Ot 5, MS 3/Ot 1, and MS 1/Ot 2.

2. *Sugar content of simple sterile hybrids in environments IBCA and IBEA.* Background of mineral supply influences the sugar content in a certain way, modifying its absolute value. The tested set of SSH was also tested on the elevated background of mineral supply in two variants—with standard (IBCA) and extended (IBEA) areas of supply. Sugar content of hybrids is summarized in Table 19.4.

TABLE 19.4 Sugar content of SSH, deviation from the average and standard in the environments IBCA and IBEA, IBCSB of NAAS, 2011–2012

Simple sterile hybrids	Area of supply					
	NBCA			NBEA		
	Sugar content (%)	Deviation from the average (%)	Deviation from St (%)	Sugar content (%)	Deviation from the average (%)	Deviation from St (%)
MS 1/Ot 2	17.7	−0.81*	3.1	18.2	1.16*	5.5
MS 1/Ot 3	17.7	0.74*	2.7	17.1	0.09	−0.7
MS 1/Ot 4	17.7	0.74*	2.7	16.8	−0.17	−2.3
MS 1/Ot 5	17.2	0.31	0.2	17.5	0.46*	1.4
MS 2/Ot 1	17.8	0.88*	3.5	17.4	0.43*	1.2
MS 2/Ot 3	15.5	−1.42*	−9.9	16.2	−0.84*	−6.1
MS 2/Ot 4	16.3	−0.62*	−5.2	16.2	−0.80*	−5.9
MS 2/Ot 5	17.1	0.14	−0.8	18.6	1.60*	8.0
MS 3/Ot 1	16.5	−0.42*	−4.1	17.8	0.83*	3.5
MS 3/Ot 2	16.6	−0.29	−3.3	17.9	0.86*	3.7
MS 3/Ot 4	16.6	−0.32	−3.5	15.3	−1.74*	−11.4
MS 3/Ot 5	17.6	0.64*	2.1	16.4	−0.60*	−4.8
MS 4/Ot 1	16.8	−0.09	−2.1	16.5	−0.47*	−4.0
MS 4/Ot 2	15.8	−1.09*	−7.9	16.6	−0.44*	−3.8
MS 4/Ot 3	16.2	−0.69*	−5.6	16.3	−0.74*	−5.6
MS 4/Ot 5	16.5	−0.42*	−4.1	16.6	−0.40*	−3.6
MS 5/Ot 1	17.3	0.81*	3.1	18.1	1.10*	5.1
MS 5/Ot 2	17.3	0.38	0.6	17.6	0.60*	2.2
MS 5/Ot 3	16.7	−0.22	−2.9	16.2	−0.74*	−5.6
MS 5/Ot 4	16.8	−0.12	−2.3	16.6	−0.44*	−3.8

Note: *—Substantially 5% of the significance level.

Analysis of Table 19.4 revealed that on the background of IBCA 5, the combinations were significantly higher than the average population value of sugar content, while on the background of IBEA—8 combinations. The rate of reaction of the studied genotypes on EA was specific: hybrids increased or lowered the sugar content, some of them showed stability. However, the range of variation characteristics was different in two environments. The difference between the highest (MS 1/Ot 2) and the lowest (MS 3/Ot 4) indexes of the sugar content was higher on the EA: it was 3.3 percent (abs. index). It was smaller in the environment IBCA—2.0 percent (hybrids MS 2/Ot 1 and MS 4/Ot 2). In relation to the standard the significant excess was 2.1...3.5 percent (IBCA) and 2.2...8.0 percent (IBEA). The high sugar content on EA was observed in hybrids MS 2/Ot 5 (18.6%), MS 5/Ot 1 (18.1%), and MS 1/Ot 2 (18.2%). Combinations of MS 2/Ot 1 (17.8%) and MS 1/Ot 2, MS 1/Ot 3, MS 1/Ot 4, 17.7 percent each, were the best on IBCA.

Determination of the proportion in the total genotypic variability revealed that it is larger than on EA (92.4%) compared to CA (83.5%). This indicates a better differentiating ability of EA than CA.

Decomposition of genotypic variance (with the help of dispersive analysis) on the effects associated with different types of gene interactions showed that nonadditive effects of genes on EA had a larger proportion (50%) compared to CA (40%) (Figures 19.3 and 19.4).

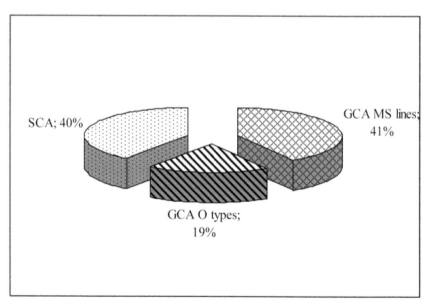

FIGURE 19.3 Genotypic variability and its share of feature of sugar content of simple sterile hybrids, 2011–2012, background IBCA.

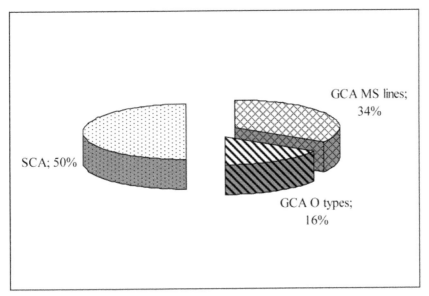

FIGURE 19.4 Genotypic variability and its share of feature of sugar content of simple sterile hybrids, 2011–2012, background IBEA

Analysis of Figures 19.3 and 19.4 showed that the effect of O types was approximately two times lower than that of MS lines: 19 vs. 41 percent (IBCA) and 16 vs. 34 percent (IBEA).

The significance of differences between SSH identified the effects of GCA and SCA components of hybridization and their expression depending on the area of supply (Tables 19.5 and 19.6). The variability of combining ability, many scholars pointed out depending on the environment. Thus, in some cases, SCA was more stable, in others—GCA [10, 11].

TABLE 19.5 The effects of GCA and SCA on sugar content of MS lines and O types, IBCA

MS lines	Effects of GCS MS lines	Effects of SCA				
		Ot 1	Ot 2	Ot 3	Ot 4	Ot 5
MS 1	0.67*	#	−0.25	0.41*	0.30	−0.47*
MS 2	−0.24*	0.72*	#	−0.84*	−0.16	0.28
MS 3	−0.09	−0.74*	0.13	#	−0.01	0.62*
MS 4	−0.56*	0.07	−0.19	0.09	#	0.03
MS 5	0.22*	0.19	0.49*	−0.22	−0.46*	#

Note: *Substantially 5% of the significance level.

As the analysis of Table 19.5 revealed, the best of SCA were MS 1 (+0.67*) and MS 5 (+0.22*). These lines possess a set of additive genes that will always influence the display of sign in F_1. In specific combinations, if the nonadditive effects of genes are expressed very well, associating with dominance and overdominance, pairs of MS 1 and Ot 3 (0.41*), MS 2, and Ot 1(0.72*), MS 3, and Ot 5 (0.62*), MS 5 and Ot 2 (0.49*) revealed themselves better. MS 4 Line did not show itself in any combination; GCA was significantly lower (−0.56*), and all hybrids with its participation were of inferior standard for 2.1…7.9 percent (Table 19.4).

On EA, MS 1 line confirmed its high degree for GCA, and also showed itself on CA. GSA of this line was significantly higher (+0.4*), and MC 5 line had a positive but nonsignificant effect (0.14) (Table 19.6).

TABLE 19.6   Effects of GCA and SCA on sugar content of MS lines and O types, IBEA

MS lines	Effects of GCS MS lines	Effects of SCA				
		Ot 1	Ot 2	Ot 3	Ot 4	Ot 5
MS 1	0.67*	#	−0.25	0.41*	0.30	−0.47*
MS 2	−0.24*	0.72*	#	−0.84*	−0.16	0.28
MS 3	−0.09	−0.74*	0.13	#	−0.01	0.62*
MS 4	−0.56*	0.07	−0.19	0.09	#	0.03
MS 5	0.22*	0.19	0.49*	−0.22	−0.46*	#

Note: *Substantially 5 percent of the significance level.

MS 4 line was characterized by a negative and significant effect on SCA both on CA and EA (Tables 19.5 and 19.6). Consequently, one can argue about the relative stability of effects of GSA, which cannot be said about effects of SCA. Thus, the components of the hybrid MS 1/Ot 3 on CA had the effect of "plus" that changed to the effect of "negative" in EA, indicating a significant expression of the effects of dominance and overdominance. Nonadditive effects of genes of MC 4 line with O types on EA showed the contrast: with the BS Ot 1 were significantly negative (−0.58*) and with Ot 3—significantly positive (+0.60*), while on CA they had no significant effect on the display of sugar content in hybrid combinations with their participation.

MS 3 line also had contrast—with Ot 1 and Ot 2 effects of SCA were positive (0.37* and 0.96*), while with Ot 4 and Ot 5—negative (−0.75* and 0.58*). On CA, MS 3 line from Ot 4 had a high positive effect of SCA. Consequently, the environment influenced the manifestation of nonadditive effects of genes, and this effect is specific for each genotype.

19.4   CONCLUSIONS

Thus, on the elevated background of fertilizing, the extended area is the environment with a good differentiating ability, as 8 hybrids were revealed, significantly exceeding average population value (compared to 5 on IBCA). Genotypic variance for IBEA was also higher (92.4%) and the range of variability was wider (3.3%) compared to IBCA—83.5 and 2.0 percent, respectively.

The best hybrids that had a stable effect on the sugar content on both backgrounds were MS 1/Ot 2, MS 2/Ot 1, MS 5/Ot 1, and MS 2/Ot 5, as evidenced by the cumulative effect of GCA and SCA. SCA was more variable in relation to the area of supply than GCA. MC 1 and MC 5 lines were distinguished as carriers of additive genes and their stable expression.

KEYWORDS

- **Combining ability**
- **Male sterility lines**
- **Simple sterile hybrids**
- **Sterility binders**
- **Sugar content**

REFERENCES

1. Savitskiy, V. F.; Genetics of Sugar Beets. Savitskiy, V. F.; Kiev: Beet Production; **1940**, *1*, 551–672 (in Russian).
2. Korneeva, M. A.; Selection and Genetic Study of Initial Populations of Sugar Beet in Order to Create Value-Combined Lines—Pollinators/Dissertation Abstract for the Degree of PhD of Biological Sciences Specialty 03.00.15—Genetics. Kiev: Institute of Molecular Biology and Genetics. **1987**, 20 p. (in Russian).
3. Balkov, I. Ya.; Heterosis of sugar beet as to its sugar content. Balkov, I. Ya.; Petrenko, V. P.; and Korneeva, M. A.; Reporter of Agricultural Science. **1986**, *10*, 55–59 (in Russian).
4. Logvinov, V. A.; Variability of sugar content in the roots of sugar beet plants and selection of plants with high sugar content. Logvinov, V. A.; Chebotar, L. L.; Increase of the Efficiency of Sugar Beet Production in the North Caucasus. Krasnodar; **1985**, 48–53 (in Russian).
5. Roik, N. V.; Combining ability of sugar beet pollinators of different genetic structure as to the elements of productivity. Roik, N. V.; Korneeva, M. A.; Encyclopedia of the Genus Beta: Biology, Genetics and Selection of Beets. Ed. Maleckij, S.; Novosibirsk: Publishing House "Sova"; **2010**, 525–541 (in Russian).
6. Korneeva, M. A.; Inheritance of sugar content by top-crossed MS hybrids of sugar beet. Korneeva, M. A.; and Vakulenko, P. I.; Sugar Beet. **2006**, *4*, 7–8 (in Ukrainian).
7. Korneeva, M. A.; The use of additive-dominant model for the evaluation of lines of sugar beet. Korneeva, M. A.; Ermantraut, E. R.; and Vlasiuk, N. V.; Methodology, Mechanization, Automation and Computerization of Researches in Arable Farming, Crop Science, Horticulture and Vegetable Growing. Kiev: Polygraph Consulting; **2007**, *9*, 164–171 (in Ukrainian).
8. Mazlumov, A. L.; Selection of sugar beet. Mazlumov, A. L.; Moscow: Publishing House "Kolos"; **1970**, 206 p. (in Russian).
9. Balkov, I. Ya.; Patterns of inheritance of sugar content and guidelines for the selection of forms with high sugar content in the selection of sugar beet. Balkov, I. Ya; Peretyatko, V. G.; Basics of Increasing of Sugar Content and Technological Properties of Sugar Beet. Kiev: All-Union Scientific Research Institute of Sugar Beet; **1986**, 70–77 (in Russian).
10. Berezhko, S. T.; Combining ability in tetraploid sugar beet and environmental conditions. Berezhko, S. T.; Genetic Studies of Sugar Beet. Kiev: All-Union Scientific Research Institute of Sugar Beet; **1986**, 75–86 (in Russian).

11. Petrenko, V. P.; The use of combining ability in the selection process. Petrenko, V. P.; Genetic Studies of Sugar Beet. Kiev: All-Union Scientific Research Institute of Sugar Beet; **1986,** 86–96 (in Russian).

CHAPTER 20

SUGAR BEET ROOT ROTS DURING VEGETATION PERIOD IN UKRAINE

NATALIYA M. ZAPOLSKA

CONTENTS

20.1   INTRODUCTION

Recent changes in agroclimate conditions in sugar beet growing have led to new problems. Different root rots and other diseases deteriorate the quality of sugar beet that is widely used in food industry, stockbreeding, and in biofuel production.

It should be emphasized that most sugar beet diseases came and still come to the territory of former USSR from the United States and Western Europe. The same holds true for other crops.

Damage of sugar beet crops with rot called "wet rot" caused by fungi *Phytophthora drechsleri* was observed in the United States as early as the 1930s [1, 2]. Twenty years later, it was already detected in the Great Britain [3]. At the end of past century, *Phytophthora* rot appeared in Ukraine on the single plants of sugar beet on the fields of Kyiv and Vinnytsia area [4].

In 1892 in Germany, Frank [5] described such diseases as heart rot and dry rot of sugar beet, caused by the fungi *Phoma betae*. Later on in 1930s, a massive manifestation of heart rot and dry rot on sugar beet crops was registered in Western Europe. Scientific reports on development of above-mentioned diseases on the territory of Ukraine were made by home researches 10 years later, in the 1940s [6].

Later on in Ukraine was determined one of the origins of heart rot and dry rot. Soil drought does not provide sugar beet plants, even partly, with water-soluble boron [7].

At the beginning of 1930s in Western Europe, a new disease of root superficial tissue of parasitic origin has revealed. Its harmfulness was significant. Later on, this disease, namely scab (*Actinomyces scabies*) appeared in Ukraine as well [8, 9].

20.2   MATERIALS AND METHODOLOGY

The field observations were carried out in different regions of Ukraine. Trials were managed using standard crop production practices described earlier [10, 11]. Root damage was valued by examination of crops during vegetation period taking into account root rots. Species composition was identified using mycological techniques [11]. Soil test was carried out by sowing its extract on Chapek agar medium using techniques of Berezova [10] and Kirai [12].

20.3   RESULTS AND DISCUSSION

The widespread is *Actinomices cretaceus Krassil* (Figure 20.1) *Actinomyces scabies Gussow and Bacillus scabies* are less spread.

FIGURE 20.1   Sugar beet root damaged with *Actinomices cretaceus* Krassil.

In recent years, *Actinomices cretaceus Krassil* often appears on new roots as early as in the middle of June, and later on under favourable condition progresses intensively in July–August. Apart from *Actinomices,* damaged roots become settled with different fungi (Figure 20.2).

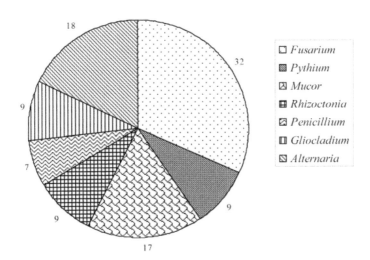

FIGURE 20.2   Mycological analysis of roots with *Actinomices cretaceus* Krassil, %.

In the composition of detected fungi, the dominant were varieties of *Fusarium* and *Pythium,* which under inevitable succession open the door for saccharolytic fungi invasion that directly causes the decomposition of tissue.

It was noted that under the abnormal weather conditions in the United States, a new rare disease of sugar beet such as crinkle has appeared [13].

On the irrigation soils of Western Europe and Middle East, dry sclerotium (*Sclerotium bataticola* Taub) manifests intensively causing negative profit in production, whereas it seldom appears in Northern Europe.

Until recently, dry sclerotium did not influence agriculture. Lately, however, growing highly productive hybrids under deteriorating soil and climatic conditions lead to increase in this disease cases [14–16]. The most affected are weak, damaged, or in a stress-state plants.

In South Spain and Northern Africa, a rather uncommon disease caused by fungi *Urophlyctis leproides* revealed on the sugar beet. The disease damages both leaves and roots, creating galls filled with mycelium [17].

Aphanomyces root rot (Figure 20.3) was first described on the beet crops in the United States, Canada, later on in Germany, Hungary, and France. It was detected on mother and factory root crops in farms of Kirovograd, Kyiv, Cherkassy, and Chernihiv areas in Ukraine, damaging 7–20 percent of roots [18–20].

FIGURE 20.3　Reveal of *Aphanomyces* root rot on sugar beet.

Pythium root rot spreads massively on sugar beet in the United States, Iran, and South Europe [21]. In Ukraine, it does not have practical importance. Nevertheless, one can find this kind of rot on the roots mostly of foreign origin. Characteristic feature of the rot is that damaged tissues become gray, then black, and of hard consistency, which greatly complicates root processing at factories (Figure 20.4).

FIGURE 20.4 Pythium root rot on sugar beet.

Rhizopus root rot is an uncommon disease in Ukraine; but recently, one can came across it more and more often. The causative agent is saprophytic micromices *Rhizopus nigricans* that features thermostability, that is, it becomes more active at the temperature of 28–45°C.

Almost everywhere in sugar beet-growing zones, storage rot (blackleg) appears every year, damaging 2–5 percent of crops [20, 22, 23]. It develops more rapidly under soil drought along with high temperature.

In recent years, under the influence of stress weather condition in Bulgaria, Hungary, India, and the United States, the manifestation of new *Fusarium* diseases have been registered on rice, corn, and tobacco crops [24].

One of the most urgent relevant problems in many countries as the United States, Belgium, Germany, India, and the Netherlands is fusarium yellows, caused with *Fusarium* [25]. Stress conditions that follow both the beginning of vegetation and of active growth period promote *Fusarium* disease development.

Fusarium diseases are observed in almost all sugar beet farms of Ukraine (Figure 20.5).

FIGURE 20.5   Fusarium yellows on sugar beet.

Marked dynamics of increase in Fusarium diseases such as *Fusarium* yellows root rot and vascular system necrosis that reveal intensively on sugar, fodder, and table beet (Figure 20.6).

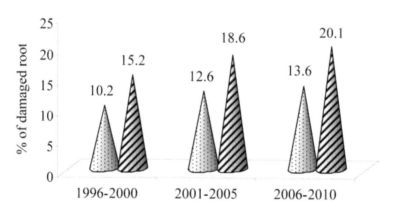

FIGURE 20.6   Dynamics of increase in Fusarium diseases.

There are mentions in the literature that sugar beet in Holland, France, England, and the United States is often affected with vascular system necrosis, the causative agent of which is soil fungi *Verticillium albo-artrum* that previously have affected rice and cotton only [26, 27].

Abnormal weather conditions provoke deformity of roots and dying off leaves before the time.

Having prolonged growing period in dry soil, in sugar beet root, mostly foreign hybrids, because of slow growth of tissue, pachypleurous cells are formed, which lose their plasticity. After rainfalls, the intensive growth of tissues is followed by skin crackling and further settling with soil fungi in it. Then deformed roots that rot rapidly are formed (Figure 20.7).

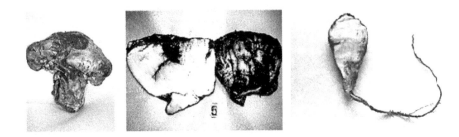

FIGURE 20.7   Deformed roots as a result of abnormal weather conditions.

Bare patches, namely brown and red root rots (Figure 20.8), are the most spread and harmful sugar beet diseases that influence the quality of beet and provoking affection of many crops that lead to increasing of rhizoctonia inoculators in soil.

FIGURE 20.8   Brown and red root rots development on sugar beet.

Researches have determined that rhizoctonia development depends greatly on weather conditions and soil humidity especially. There is a direct relation between soil humidity and quantity of inoculators in it. Thus, increasing soil humidity by 1 percent is followed with rise in number of rhizoctonia fungi germs (i.e., spores) as much as 10 percent and more (Figure 20.9).

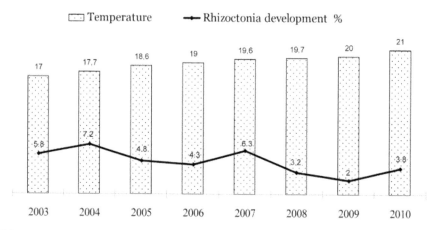

FIGURE 20.9 *Rhizoctonia* development depending on temperature in Ukraine.

A biological peculiarity of the fungi should be mentioned, which is the highest radial velocity of growth when compared with other fungi.

One of the most powerful factors forming phytosanitary state of agrocenosis and defining its infection potential is soil.

It should be emphasized that certain groups of micromyces cenosis are selected by crops through root exudates. Therefore, variety of pathogens, saprophytes, and toxin-producing microflora rely greatly on crops.

In recent years, because of the intensive use of soil in agrocenosis associated with long-term growing of the same crops, factor of soil supressivity, that is, an ability of soil to suppress pathogenic microbiota and epiphytoty is decreasing.

The changes in structure of crops have greatly deteriorated phytosanitary state of soils because crops often grow at the same place for some years in a row. This leads to worsening of not only soil microbiological activity but also of its fungistatical properties (Figure 20.10).

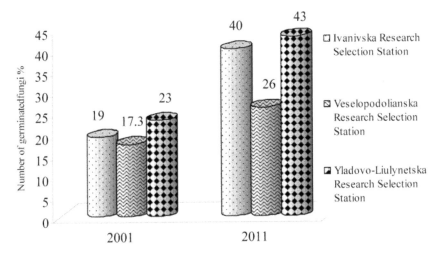

FIGURE 20.10   Changing of soil fungistatical properties, 2010–2011.

Fungistatical ability is common for all soil types. It is formed by plants and prevents not only contamination but also pathogenic fungi growth, which results in reduction of soil infection.

It should be marked that typical soils have different extent of fungistatical activity concerning reproductive organs of many fungi that are important in keeping their vital activity.

Thus, an increase in share grain crops promotes not only one-way carrying out of nutrients but also increasing in phytotoxic fungi (Table 20.1).

TABLE 20.1   Share of different agricultural crops, %

Agricultural crops	2000	2010
Cereals (without corn)	52	49
Sunflower	13	18
Corn	12	12
Permanent grasses	9	4
Buckwheat	3	1.2
Rape	2	4
Soya	0.3	4
Sugar beet	6	3
Pea	2	1.5
Other crops	4	4.5

Mycological tests of soils have showed that among the fungi *Penicillium* species a great share of selected varieties made up toxin producing ones: *Penicillium cyaco-fulvum, P. rubrum, P. purpurogennum*. At the same time, the number of toxic species such as *Aspergillus, Cladospopium, Stemphilium, Alternaria, and Mucorales* has increased. Besides, the number of fungi-antagonists *Trichoderma and Botrytis* is decreased.

The plants resistance to diseases on contaminated soils decreases. It is especially true if hybrids option for sowing is improper to a zone and followed by unfavorable weather conditions. Thus, early affection of roots with causative agents of deceases is common.

Popkova determined a correlation between potato periderm thicknesses and its resistance to affection.

As early as 1990s, Japanese scientists noted that sugar beet cell walls, having rapidly increasing weight, were less thick and receptive to destruction with fungi [28].

The diverse origin hybrid cells examination has shown the differences in the size of cell walls and intercellulars. One can see the differences in the cell density in the first 10 layers (Figures 20.11 and 20.12).

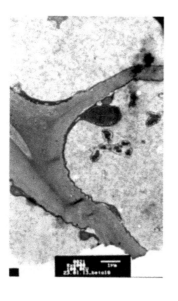 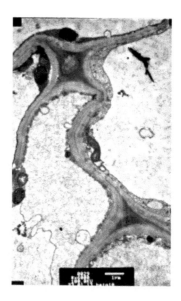

A   B

FIGURE 20.11 Size of cell walls and intercellulars of sugar beet hybrids of different origin: (A)—home hybrid; (B)—foreign hybrid.

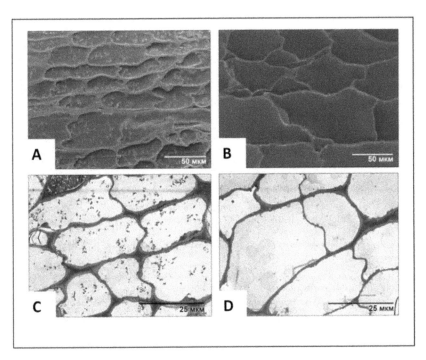

FIGURE 20.12 Ultrastructural analysis of the cells of the first 10 layers concerning sugar beet root (neck area): (A), (C)—cell size in the top area of sugar beet root (home hybrid); (B), (D)—thickness of cell walls and intercellulars (foreign hybrid).

20.4 CONCLUSIONS

Thus, the issue of sugar beet diseases and rots especially at the vegetation period appeared because of the complex breakage of environment—plant—pathogen chain (environment here is changing climate and agrotechnic conditions for sugar beet growing). Directed efforts to increase the yield and sugar content change the morphologic properties of plant. Under described conditions herein above, saprophytes such as *Fusarium* fungi, *Alternaria, Stemphylium*, and mukors evolve to pathogens. All that has resulted in such problem as sugar beet root rots during vegetation period in Ukraine.

KEYWORDS

- Aphanomyces root rot
- Fusarium root rot
- Fusarium yellows
- Pythium root rot
- Rhizopus root rot
- Scab (actinomyces scabies)

REFERENCES

1. Stirrup, H. H.; Sugar beet diseases. *Ann Appl Bio.* **1939**, *26(4),* 402–408.
2. Tomkins, C. M.; Richards, B. I.; Tucker, C. M.; and Gardner, M. W.; Phytophtora rot of sugar beet. *J. Agric. Res.* **1936**, *52(3),* 205–216.
3. Gates, L. F.; and Hull, R.; Experiments on black leg disease of sugar-beet seedlings. *Ann. Appl. Biol.* **1954**, *41(4),* 541–561.
4. Sabluk, V. T.; Shendryk, R. Y.; and Zapolska, N. M.; Pests and Diseases of Sugar Beet. Kyiv: Kolobig; **2005**, 448 p. (in Ukrainian).
5. Frank, A. B.; Über *Phoma betae*, einen neuen parasitischen Pilz, welcher die Zuckerrüben zerstört. *Zeitschrift der Vereines für die Rübenzuckerindustrie des Deutschen Reiches.* **1892**, Band. 42, S. 904–916 p.
6. Zavgorodniy, F. I.; Effect of Boron on Growth, Yield and Composition of Sugar Beet. Collected Papers of Kyiv: Agricultural Institute; **1940**, *2,* 51–64 p. (in Russian).
7. Petrik, M. P.; Patterns of Heart Rot and Dry Rot of Sugar Beet Development and Measures of the Disease Control in the Western Regions of the USSR Thesis Abstract Kyiv. **1984**, 19 p. (in Ukrainian)
8. Büttner, G.; Pfähler, B.; and Märländer, B.; Greenhouse and Field Techniques for Testing Sugar Beet for Resistance to Rhizoctonia Root and Crown Rot. Plant Breeding; **2004**, *123(2),* 158–166 p.
9. Mirchink, T. G.; Soil Mycology. Moskow: "MSU Press"; **1988**, 220 p. (in Russian).
10. Berezova, E. F.; Microflora of Root Systems of Plants and Methods of its Study. Proceedings of the Institute for Microbiology; **1951**, *12,* 39–55 p.
11. Bilai, V. I.; Methods of Experimental Mycology. Kyiv: "Naukova Dumka"; **1982**, 562 p.
12. Kirai, Z.; and Klement, V.; Methods of Phytopathology. Moskow: "Kolos"; **1974**, 343 p.
13. Coons, G. H. D.; Stewart, H. W.; and Bockstahler, C. L.; Schneider. Incidence of Savoy in Relation to the Variety of Sugar Beets and to the Proximity of the Winter Vector *Piesma Cinerea.* Plant Disease Reporter; **1958**, *42,* 502–511 p.
14. Shendryk, R. Y.; and Zapolska, N. M.; Forecast of disease development in Ukraine in 2009. Phytosanitary State Agrocenosis Forecast for Ukraine and Advice on Plant Protection in 2009. Ministry of Agriculture. State Inspectorate for Plant Protection; **2009**, 76–80 p. (in Ukrainian).
15. Shendryk, R. Y.; and Zapolska, N. M.; Forecast of the diseases in 2010. Kyiv. Phytosanitary state agrocenosis forecast for Ukraine and advice on plant protection in 2010. Ministry of Agriculture. State Inspectorate for Plant Protection; **2010**, 79–81 p. (in Ukrainian).
16. Shendryk, R. Y.; and Zapolska, N. M.; Forecast of the disease in 2011. Phytosanitary state agrocenosis forecast for Ukraine and advice on plant protection for 2011. Ministry of Agriculture. State Inspectorate for Plant Protection; **2011**, 89–92 p. (in Ukrainian).
17. Whitney, E. D.; The First Confirmable Occurrence of Urophlyctis Leproides on Sugar Beet in North America. Plant Disease Reporter; **1991**, *55,* 30–32 p.

18. Shendryk, R. Y.; and Zapolska, N. M.; Pests and diseases of sugar beet. Phytosanitary state agrocenosis forecast for Ukraine and advice on plant protection in 2001. Ministry of Agriculture. State Inspectorate for Plant Protection in 2001. **2001,** 52–53 p. (in Ukrainian).

19. Shendryk, R. Y.; and Zapolska, N. M.; Phytosanitary state agrocenosis forecast for Ukraine and advice on plant protection in 2006. Ministry of Agriculture. State Inspectorate for Plant Protection in 2006. **2006,** 58–61 p (in Ukrainian).

20. Shendryk, R. Y.; Phytosanitary state agrocenosis forecast for Ukraine and advice on plant protection in 2007. Ministry of Agriculture. State Inspectorate for Plant Protection; **2007,** *60–64,* (in Ukrainian).

21. Windels, C. E.; and Jones, R. K.; Seedling and Root Diseases of Sugarbeets. Univ. Minnesota Ext. Serv. AG-FO-3702. **1989,** 8 p.

22. Shendryk, R. Y.; and Zapolska, N. M.; Sugar beet diseases. Phytosanitary state agrocenosis forecast for Ukraine and advice on plant protection in 2008. Ministry of Agriculture. State Inspectorate for Plant Protection in 2008. **2008,** C. 66–73 p. (in Ukrainian).

23. Shendryk, R. Y.; and Zapolska, N. M.; Forecast for the diseases in 2009. Phytosanitary state agrocenosis forecast for Ukraine and advice on plant protection in 2009. Ministry of Agriculture. State Inspectorate for Plant Protection in 2009. **2009,** C.76–80 p. (in Ukrainian).

24. Karadjova, L. V.; Crops Fusarium Diseases. Kishinev: "Shtiinca"; **1989,** 253 p. (in Russian).

25. Martyn, R. D.; Rush, C. M.; Biles, C. L.; and Baker, E. N.; Etiology of a root rot diseases of sugar beet in Texas. Plant Disease. **1989,** *73,* 879–884 p.

26. Franc, G. D.; Harveson, R. M.; Kerr, E. D.; and Jacobsen, B. L.; Disease Management. Sugarbeet Production Guide. Lincoln, NE: Regional Bulletin EC University of Nebraska Cooperative Extension; **2001,** 131–160 p.

27. Schneider, C. L.; and Whitney, E. D.; Fusarium Yellows. Compendium of Beet Diseases and Insects. Ed. Whitney, E. D.; and Duffus, J. E.; Sant Paul, Minnesota: APS Press; **1986,** 18 p.

28. Uchino, H.; Watanabe, H.; and Kanzawa, K.; Controlling Root Diseases of Sugar Beet by Applying Azoxystrobin. Proceedings of the Japanese society of sugar beet technologists: Papers presented at the 37th meeting in Sapporo. Tokyo: **1997,** *39,* 73–79 p. (in Japan).

CHAPTER 21

CYTOGENETIC AFTER-EFFECTS OF MUTAGEN SOIL CONTAMINATION WITH EMISSIONS OF BURSHTYNSKA THERMAL POWER STATION

RUSLAN A. YAKYMCHUK

CONTENTS

21.1　INTRODUCTION

Current ecological situation in Ukraine is characterized by a considerable anthropo-genic burden in the form of mutagens of physical and chemical nature. Total xenobi-otic contamination of atmospheric air, soil, drinking water, and foodstuff was caused by genetically explained pathology; the latter leads to birth defects of the development and cytogenetic disorders in gametal and somatic cells [1, 2].

One of the main contaminants of the biosphere of industrial cities is heat power engineering; its share is 27–32 percent of the total volume of contaminating emissions; recently, they have been 100 million (metric) tons [3]. Thermal electric stations emit mutagens into free air, including heavy metals, radioisotopes, benz(a)pyrene, dioxins, and other chemical carcinogens [4]. The analysis of the contaminating sources con-firms that the major contaminant of the biosphere with heavy metals, which comes in an aeroindustrial way, is thermal power station (TPS); their emissions contain Cu, Zn, Pb, As, Hg, Ni, V, Cr, and Al [5]. Despite the fact that they play an active role in biochemical reactions in trace amounts, they are toxic and capable of reducing natural resistance of biological objects to biotic and abiotic factors of the environment when they are in large amounts [6]. The most important depositor of heavy metals both in natural and artificially created ecosystems is soil. Heavy metals stay much longer in the soils than in other natural bodies. Thus, conditionally the soil contamination with heavy metals can be considered "everlasting" (metals are not destroyed; they shift into another form of existence, including the composition of salts, oxides, and metalor-ganic compounds). The period of semiremoval for copper is 310–1,500 years, for zinc is 70–510 years, for lead is 740–5,900 years, and for cadmium is 13–110 years [7].

TPS as compared with properly functioning nuclear power stations are more dan-gerous sources of radiation contamination, because additional radiation doses, which people who live near TES receive, are 40 times higher than those caused by nuclear reactors [8–10]. Cytogenetic studies of the lymphocytes of peripheral blood of the workers of Kemerovska TPS (Siberia) showed certain increase of chromosome aber-ration frequency when compared to a control group [11]. The highest chromosome aberration frequency was typical for the workers of the chemical workshops, and was lowest for the workers of the repair shops.

The coal used at TPS contains a relatively small quantity of primary radionuclides: on average, potassium—40–50 Bq/kg, uranium-238 and thorium-232—20 Bq/kg, and its contribution to a radiation dose of people is relatively small. Disproportionation/redistribution of radionuclides from the ground into a biosphere takes place during coal getting and burning, and using coal ash for building materials, which explains radiation increase of the population [8].

Solving ecological problems concerning environmental pollution with TPS emis-sions requires the development and application of the monitoring system of different kinds in the areas adjacent to stationary sources of thermal power engineering. Most of the research is aimed at studying the accumulation level of heavy metals, [12] ra-dioisotopes [13] in the soil, water reservoirs, and plants depending on the distance to the pollution source. Genetic studies are very important in the system of biological monitoring, as they make possible the evaluation of consequences of the simultaneous effect of several stress factors for the consecutive generations at cellular and molecular levels [1, 14, 15]. To systematically and efficiently remove or reduce harmful anthropogenic after-effects of the polluted areas, it is necessary to carry out a systematic diagnostics of a soil surface condition that records the main tendencies of long-term pollution processes [16].

With this end in view, mutagen activity of soil contaminants of the area adjacent to Burshtynska TPS as to the frequency and spectrum of chromosome aberrations was studied.

21.2 MATERIALS AND METHODOLOGY

To determine mutagen activity of soil pollutants of the area adjacent to Burshtynska TPS (Ivano-Frankivsk region), a cytogenetic analysis of meristem cells of primary roots of winter wheat plantlets (*Triticum aestivum* L.) of 'Albatross odeskyi' and 'Zymoiarka' cultivars was made. Seeds were kept for 40 min in moist soil, taken from the place, which is 1, 3, 4, and 12 km from a pollution source on the axis of air mass transfer and near ash-disposal area No 1. Soil sampling was done in accordance with standard techniques [17] and the requirements of national standard No 17.04.3.01.83, No 17.4.4.02.84.

Taking into consideration the fact that the soils of Poltava region were not polluted with radionuclides as a result of Chernobyl catastrophe and heavy metal content was much lower than permissible concentrations, [16] the soil sampled from the area near Svatky, Hadiach district, Poltava region was taken as control. The analysis aimed at identifying heavy metal content in soil samples was made in the department of agroecology and analytical research of the NSC 'Institute of arable farming of the National Academy of Agrarian Sciences of Ukraine.' The concentration of mobile forms of lead in the soil samples taken from the area adjacent to TPS and ash-disposable area No 1 was 1.0–1.5 maximum allowable concentrations and exceeded control level by 1.6–3.3 times.

Seeds were germinated at 24–26°C. Primary roots (0.8–1.0 cm long) were kept in 'vinegar alcohol' and exposed to maceration in muriatic acid solution. Crushed temporary preparations were made from the roots of apical meristem. When mitosis disorder frequency and chromosome aberration were being determined, the cells in an anaphase and early telophase were taken into account. The sample comprised of at least 1,000 cells for each variant.

21.3 RESULTS AND DISCUSSIONS

21.3.1 FREQUENCY OF CHROMOSOME ABERRATIONS

Ukraine holds one of the first places in Europe as to the number of harmful emissions. In Ivano-Frankivsk region, their amounts are 0.1–0.4 t/ha per year, which makes the region equal to highly industrialized regions of the south-east area of the country as to environmental pollution [18]. The process of coal burning is the main source of heavy metals coming into a biosphere and an additional factor of the increase of natural radiation background. Resultant concentration of metals per fuel ton is 500 g. Considering the fact that during the whole history of mankind, 130 billion tons of coal were burnt and [19] 65 million tons of metals were emitted and additionally added to circulation. Coal, slag, and ash resulting from pyrolysis contain 7–10 times more of primary radionuclides than soil (potassium—40–400 *Bq/kg*, uranium-238 and -235—150 *Bq/kg*) [8]. Volatile ash is taken up by hot gases, and it partially enters the atmosphere and is absorbed by soil surface [13]. Therefore, it can be assumed that the main mutagens that contaminate adjacent areas of Burshtynska TPS are heavy metals and natural radioisotopes. A recently developed system of ecologic-genetic monitoring envisages bioindication of mutagens of the environment with the help of a cytogenetic analysis. Its advantage is the feasibility of getting mutagen estimation of natural environment regardless of the composition of contaminating substances [15]. To perform monitoring of real environmental pollution with potentially harmful compounds (in genetic sense), plant test systems, to which soft wheat was included, are the most appropriate (*T. aestivum*) [20, 21].

The analysis of chromosome aberration frequency in meristem cells of winter wheat roots, whose seeds were germinated in soil samples taken near Burshtynska TPS, showed its growth by 1.9–2.5 times (as compared with control) for cultivar 'Albatross odeskyi' and by 1.5–2.4 times for 'Zymoiarka' cultivar (Table 21.1). The largest number of chromosome aberrations was recorded with the contamination effect of the soil in 3–6 km zone. There was no frequency excess of aberrated cells of plantlets, which were affected by soil factors when compared to control level (a soil sample was taken at a distance of 1–12 km from the source of pollution). However, soil pollution in 1 km area from TPS caused 1.47 ± 0.10 percent cells with chromosome disorders in meristem roots of Zymoiarka plantlets, which certainly exceeded control indices ($0.75 \pm 0.25\%$). The analyses of toxic element content in the soil near Burshtynska TPS (according to the data of other authors [13, 22]) showed their maximal accumulation in the area of 3–8 km from the source of contamination toward prevailing winds. Lead, even in small concentrations, is one of the most dangerous pollutants of the environment, which can depress reparatory processes, enhancing in turn genetic after-effects of low radiation doses [23–25]. Based on the results received, it can be assumed that an increased level of chromosome aberrations caused by soil contaminants is the outcome of a synergistic effect of at least two mutagen factors—lead and primary radionuclides.

The increase of a cytogenetic disorder level was observed with the effect of soil pollutants in the area of ash-disposal No 1. The frequency of the cells with chromosome disorders was at the level of 1.63 ± 0.38 percent in wheat of 'Albatross odeskyi' cultivar and 1.32 ± 0.35 percent in 'Zymoiarka,' which exceeded control level by 3.0 and 1.8 times, respectively.

TABLE 21.1 Frequency of chromosome aberrations in winter wheat when mutagen soil pollution occurs in the area of Burshtynska TPS

Sampling location	Studied		Mitosis with disorders and chromosome aberrations	
	Roots	Mitosis anatelophase	Pieces	(%)
'Albatross odesskyi'				
Svatky Poltava region (control)	24	1291	7	0.54 ± 0.22
1 km from TPS	18	1195	12	1.00 ± 0,29
3 km from TPS	18	1282	13	1.01 ± 0.28
5 km from TPS	24	1248	17	1.36 ± 0.33*
12 km from TPS	18	1328	15	1.12 ± 0.29
Ash-disposal area no 1	15	1101	18	1.63 ± 0.38*
'Zymoiarka'				
Svatky Poltava region (control)	24	1200	9	0.75 ± 0,25
1 km from TPS	15	1496	22	1.47 ± 0.10*
3 km from TPS	16	1047	19	1.82 ± 0.41*
5 km from TPS	18	1350	23	1.70 ± 0.35*
12 km from TPS	18	1098	12	1.10 ± 0.31
Ash-disposal area no 1	15	1063	14	1.32 ± 0.35

Note: *Difference with control statistically certain when $P < 0.05$

21.3.2 SPECTRUM OF CHROMOSOME ABERRATIONS

The spectrum of chromosome aberrations that were induced by emissions of Burshtynska TPS, absorbed by the soil in 3 and 5 km zone, included mostly singular and pair/double acentric fragments and dicentric bridges (Figures 21.1 and 21.2). Types of chromosome disorders with the effect of the soil taken at a distance of 1–12 km from the contamination source were presented by singular and double/pair fragments and chromatic bridges (Table 21.2). Cells with pair fragments and chromosome bridges occurred very rarely. The frequency of the cells with plural aberrations was 0.07–0.29 percent, and their emergence was due to the factors of soil pollution in a 3-km zone (Figure 21.3).

Along with the mentioned chromosome aberrations in the cells of the samples under study, contrary to control, micronuclei and lagging chromosomes were recorded (Figure 21.4). The latter are the indicators of mitosis anomaly, and they confirm the aneugenetic effect of contaminating factors [26, 27]. The frequency of their fixation ranged from 0.08 to 0.19 percent. The germination of wheat seeds ('Zymoiarka' cultivar) in the soil taken from a 1-km zone resulted in the appearance of the cells with tripolar mitosis.

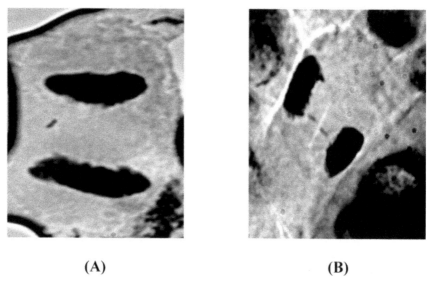

(A)   (B)

FIGURE 21.1   Singular (A) and pair/double (B) acentric fragments.

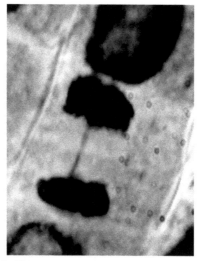

FIGURE 21.2   Dicentric **bridge.**

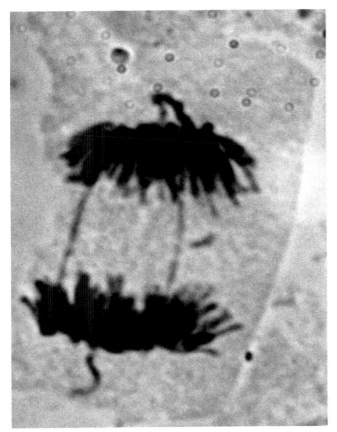

FIGURE 21.3   Cell with plural aberrations.

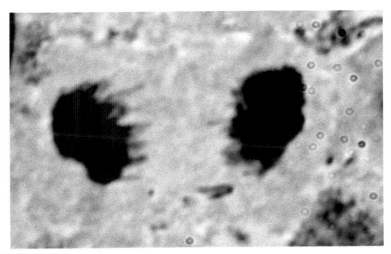

FIGURE 21.4   Lagging chromosome.

TABLE 21.2 Spectrum of chromosome aberrations in winter wheat when mutagen soil pollution occurs in the area of Burshtynska TPS

Sampling location	Spectrum of mitosis disorders and chromosome aberrations									
	Fragments		Bridges		Bridges + fragments		Micro-nuclei		Lagging chromosomes	
	pcs	%	pcs	%	pcs	%	pcs	%	pcs	%
'Albatross odesskyi'										
Svatky Poltava region (control)	1	0.08	6	0.47	0	0.00	0	0.00	0	0.00
1 km from TPS	3	0.25	7	0.59	0	0.00	2	0.17	0	0.00
3 km from TPS	5	0.39	5	0.39	1	0.08	1	0.08	1	0.08
5 km from TPS	4	0.32	13	1.04	0	0.00	0	0.00	0	0.00
12 km from TPS	9	0.68	4	0.30	0	0.00	0	0.00	2	0.15
Ash-disposal area no 1	6	0.55	11	1.00	0	0.00	0	0.00	1	0.09
'Zymoiarka'										
Svatky Poltava region. (control)	5	0.42	4	0.33	0	0.00	0	0.00	0	0.00
1 km from TPS	5	0.33	13	0.87	1	0.07	1	0.07	2	0.13
3 km from TPS	5	0.48	10	0.96	3	0.29	0	0.00	2	0.19
5 km from TPS	11	0.82	9	0.67	0	0.00	2	0.15	1	0.07
12 km from TPS	6	0.55	5	0.46	0	0.00	0	0.00	1	0.09
Ash-disposal area no 1	8	0.75	6	0.56	0	0.00	0	0.00	0	0.00

A high level of chromosome alteration, caused by soil pollution near ash-disposable No 1 of Burshtynska TPS, did not show a wide spectrum of their types and contained mostly acentric fragments and dicentric bridges. In meristem cells of wheat plantlets of 'Albatross odeskyi,' contrary to 'Zymoiarka' cultivar, a dominating aberration type was pair fragments and chromosome bridges. An anaphase cell with a lagging chromosome was identified among them.

21.4 CONCLUSIONS

A complex effect of soil pollutants of the areas adjacent to Burshtynska TPS and ash-disposal area No 1 therefore causes the increase of the cytogenetic disorder level in meristem root cells of winter wheat plantlets by 1.5–2.5 and 1.8–3.0 times, respectively. The highest chromosome aberration frequency was recorded with the mutagen effect of the soil taken at a distance of 3–5 km from the contamination source. Their spectrum is mostly presented by singular and pair fragments, and dicentric bridges. Harmful emissions of Burshtynska TPS, absorbed by the soil, result in mitosis anomaly, which is seen in blocking the threads of division spindle and the emergence of the cells with lagging chromosomes. The increased mutation level, found in the vicinity of Burshtynska TPS, confirms the fact of potential negative genetic after-effects of the emissions from the stationary sources of thermal power engineering on the organisms; and it requires further implementation of the monitoring of the areas adjacent to TPS of Ukraine.

KEYWORDS

- **Chromosome aberrations**
- **Genetic consequences**
- **Mitosis disorder**
- **Mutagens**
- **Triticum aestivum L.**

REFERENCE

1. Mamedova, A. O.; Bioindication quality of the environment on the basis of mutation and modification variability of plant. *Cytol Genet.* **2009,** *43(2),* 61–64 (in Russian).
2. Suskov, I. I.; Kuzmin, N. S.; Suskova, V. S.; et al., Individual features of trans generational genomic instability in children of liquidators of the chernobyl accident (cytogenetic and immunogenic indicators). *Rad Biol. Radioecol.* **2008,** *48(3),* 278–286 (in Russian).
3. Lykholat, Y. V.; Hrhoryuk, I. P.; Balalaeva, D. C.; et al., Accumulation of heavy metals in the organs of flowers and ornamental plants under different environmental conditions. *Rep. NAS Ukraine.* **2007,** *7,* 203–207 (in Ukrainian).
4. Timoshevsky, V. A.; and Nazarenko, S. A.; Interphase cytogenetic in the evaluation of genomic mutations in somatic cells. *Genet.* **2005,** *41(1),* 5–16 (in Russian).
5. Kornelyuk, N. M.; and Myslyuk, A. A.; Anthropogenic factors aerotechnogenic pollution Cherkassy heavy metal. *Ecology Environ. Safety.* **2007,** *4,* 48–53 (in Ukrainian).

6. Nies, D. H.; Microbial heavy-metal resistance. *Appl. Microbiol. Biotechnol.* **1999,** *51,* 730–750.

7. Orlov, D. S.; Sadovnikova, L. K.; and Lozanovskaya, I. N.; Ecology and Conservation of the Biosphere by Chemical Pollution. Moscow: Higher School; **2002,** 146–157 p. (in Russian).

8. Zaharchenko, M. P.; Havinson, V. H.; Onikienko, S. B.; and Novozhylov, G. N.; Radiation, Ecology, Health. Gumanistika: St. Petersburg; **2003,** 336 p. (in Russian).

9. Klimenko, M. O.; Pryshchepa, A. M.; and Voznyuk, N. M.; Environmental Monitoring: a Textbook [for College Students]. Kyiv: Publishing House "Academy"; **2006,** 360 p. (in Ukrainian).

10. Nakonechniy, Y. Y.; Environmental Monitoring Several Regions of Ukraine and the Use of Natural Sorbents for Improvement of Ecological Situation: Dissertation for the Degree of Candidate of Sciences: Spec. 21. 06. 01 "Techno Security State". Kyiv; **2000,** 17 p. (in Ukrainian).

11. Savchenko, Y. A.; Druzhinin, V. G.; Minin, V. I.; et al., Cytogenetic analysis of genotoxic effects in workers of heat and power production. *Genet.* **2008,** *44(6),* 857–862 (in Russian).

12. Shkvar, A. I.; Impact of anthropogenic impact Burshtyns'ka TPP surrounding area. Scientific Bulletin of National University of Life and Environmental Sciences of Ukraine. **2011,** *162(2),* 106–110 p. (in Ukrainian).

13. Grabowski, V.; and Bratash, A.; Assessment of radioactive emissions Dobrotvir TPP (Lviv region.) and their impact on the environment. *Electron. Inf. Technol.* **2011,** *1,* 166–175 (in Ukrainian).

14. Boguslavska, L. V.; Shupranova, L. V.; and Vinnychenko, O. N.; Cytogenetic activity of the root meristem cells of maize plants by separate and joint action of heavy metal ions. *Bull. Ukrainian Soc. Genet. Breeders.* **2009,** *7(1),* 10–16 (in Ukrainian).

15. Bodnar, L. S.; Matsyah, A. V.; and Belyaev, V. V.; Monitoring Genetoxicological Contamination of Some Environmental Factors. Genetics and Breeding in Ukraine Turn of the Millennium. Kyiv: Logos; **2001,** 219–225 p. (in Ukrainian).

16. Program of Environmental Protection, Sustainable Use of Natural Resources and Environmental Policy Based on Regional Priorities, Poltava Region Until 2010 in the New Edition. Poltava: Poltava Writer; **2007,** 162 p. (in Ukrainian).

17. Becker, A. A.; and Agaev, T. B.; Protection and Control of Environmental Pollution. Leningrad: Gidrometeoizdat; **1989,** 286 p. (in Russian).

18. Korsun, S. G.; Davydyuk, G. V.; Kozeretska, I. A.; et al., Influence of heavy metals on the performance, quality and genetic activity of soy. *Agroecol. J.* **2005,** *3,* 37–40 (in Ukrainian).

19. Muravyov, A. T.; Karryev, B. B.; and Lyandeberg, A. R.; Assessing the environmental condition of the soil. Krisma: St. Petersburg; **2000,** 164 p. (in Russian).

20. Bittueva, M. M.; Abilev, S. K.; and Tarasov, V. A.; The effectiveness of prediction of carcinogenic activity of chemical compounds with allowance for soybean somatic mutations *Glycine max* (L.) Merrill. *Genet.* **2007,** *43(1),* 78–87 (in Russian).

21. Morgun, V. V.; and Logvynenko, V. F.; Mutation Breeding of Wheat [Monograph]. Kyiv: Naukova Dumka; **1995,** 624 p. (in Russian).

22. Shvets, L. S.; Bioindication intensity of pollution in terms of fertility of pollen grains of various plants. *Adv. Biol. Med.* **2011,** *17(1),* 40–44 (in Ukrainian).

23. Zaichkina, S. I.; Rozanov, O. M.; Aptikaeva, G. F.; et al., The combined effect of heavy metal salts, chronic and acute exposure to the γ-value cytogenetic damage the bone marrow cells of mice and rats. *Rad Biol. Radioecol.* **2001,** *41(5),* 514–518 (in Russian).

24. Musienko, M. M.; and Kosik, O. I.; Effect of lead on environmental and physiological parameters of plants. *Bull. KNU. Taras Shevchenko, August. Biol.* **2002,** *36/37,* 37–40 (in Ukrainian).

25. Yagunov, A. S.; Tokalov, S. V.; Potyavina, E. V.; et al., Combined effects of long-acting γ-radiation and heavy metal ions in the blood system of rats. *Rad. Biol. Radioecol.* **2006,** *46(1),* 23–26 (in Russian).

26. Serduk, A. M.; Timchenko, O. I.; Lynchak, A. V.; et al., Health in Ukraine: the impact of genetic processes. *J. Med. Sci. Ukraine.* **2007,** *13(1),* 78–92 (in Ukrainian).

27. Maffei, F.; Fimognari, C.; Castelli, E.; et al., Increased cytogenetic damage detected by FISH analysis on micronuclei in peripheral lymphocytes from alcoholics. Mutagenesis. **2000,** *15,* 517–523 p.

CHAPTER 22

THE EFFICIENCY OF THE USE OF CROP PRODUCE FOR ALTERNATIVE FUEL PRODUCTION

ANDREI V. DORONIN

CONTENTS

22.1   INTRODUCTION

The influence of internal and external factors on the competitiveness of domestic agricultural enterprises requires the manufacturers to intensely form competitive advantages of products and ensure its competitive position in the market. However, in view of an urgent problem of providing our country with price-affordable energy carriers, it becomes appropriate to speed up the alternative fuel production.

Sugar and starch crops are used worldwide for the bioethanol production, including the output product of sugar beet processing. Sugar beet industry has always played an important role in Ukraine's economy and social development of society. Alongside with sugar production, and as the country needs energy carriers very much, it is quite relevant to use sugar beets and the output product of sugar beet processing for the production of bioethanol as an alternative fuel.

A considerable contribution to the development of the competitiveness of sugar beet companies, including the explanation of the ways of its enhancement, was made by Varchenko [1], Zayets [2], Kaletnik [3], Kodenska [4], Royik [5], Sabluk [6], Spychak [7], Bondar, Fursa, Yarchuk, and others. The essence of innovative technologies of the alcohol industry is studied by Shyian et al. [8] and others.

The goal of our research is to develop practical recommendations, which help to ensure the competitive manufacturing of bioethanol using the output product of sugar beet processing.

The developed world and European Union countries make great efforts to replace traditional kinds of fuel with biofuels. Thus, the EU directive Renewable Energy Directive (RED) 2009/28/EU establishes the use of 10 percent of renewable energy in transport and 20 percent of renewable energy in the structure of gross energy consumption by 2020 [9] as mandatory parameters.

The Law of Ukraine "On Alternative Fuels" [10] had envisaged that from 2013, it was recommended to add at least (not less than) 5 percent of bioethanol into gasoline; this rule has become mandatory in 2014–2015. From 2016, a required content of bioethanol in motor gasoline, which is produced and/or sold on the territory of Ukraine, will not be less than 7 percent.

Ukraine is obliged to consider the European standards, concerning the use of biofuels in the context of its entry into the European Energy Community. Therefore, the country has the obligation to bring the biological component in motor fuel up to 10 percent by the year of 2020.

22.2   MATERIALS AND METHODS

The data received from the State Statistics Department of Ukraine and the author's own calculations were used in the process of writing an article. Prices and production costs were translated at the official exchange rate of hryvnia to the US dollar, which is established by the National Bank of Ukraine for an appropriate period. The prices were given excluding value-added tax (VAT—20%), budget subsidies, and surcharges. The coefficients of ratio between the cost of production of sugar beet processing and prices of sweet roots were calculated based on the fact that the price of sugar beets is 1.0. The methods of system analysis and logic generalization were used to study the experience of bioethanol production from crop produce; the comparative analysis was used in the process of analyzing the statistical information; the economic-mathematical modeling was used to develop a polynomial model, which describes the level of sugar beet productivity in Ukraine; the monographic method helped to substantiate the necessity of diversification of the products of sugar beet industry, and induction and deduction facilitated the summarizing of the results of the research; an abstract logic method was applied to make conclusions and proposals.

22.3   RESULTS AND DISCUSSION

Bioethanol production is possible at the ethyl and sugar factories reequipped for this production. The manufacture in several areas is possible at sugar processing factories, namely the workshop producing ethanol is mounted—the plant produces sugar by the traditional technology, and it manufactures bioethanol using the products of sweet root processing, or only the bioethanol production from sugar beet (crude juice) is planned. In addition, the bioethanol production is possible when starch crops such as wheat and corn are used; the products of their processing are used for human nutrition.

Over the past 12 years, sugar beet, wheat, and corn production had undergone significant changes for the betterment in Ukraine (Table 22.1) [11].

TABLE 22.1   The indices of sugar beet, wheat, corn production in Ukraine in the years 2000–2012

Crop production indices	2000	2005	2010	2011	2012	%, 2012 in comparison with	
						2000	2011
Sugar beets							
Harvested area, thousand hectares	747.0	623.3	492.0	515.8	448.9	60.1	87.0
Production, thousand tons	13198.8	15467.8	13749.2	18740.5	18438.9	139.7	98.4
Yield, thousand hectares	17.67	24.82	27.95	36.33	41.08	232.5	113.1
Wheat							
Harvested area, thousand hectares	5161.6	6571.0	6284.1	6657.3	5629.7	109.1	84.6
Production, thousand tons	10197.0	18699.2	16851.3	22323.6	15762.6	154.6	70.6
Yield, thousand tons	1.98	2.85	2.68	3.35	2.80	141.4	83.6
Corn							
Harvested area, thousand hectares	1278.8	1659.5	2647.6	3543.7	4371.9	341.9	123.4
Production, thousand tons	3848.1	7166.6	11,953.0	22,837.8	20,961.3	544.7	91.8
Yield, thousand tons	3.01	4.32	4.51	6.44	4.79	159.1	74.4

Following the technology elements of sugar beet cultivation, the main ones are plant nutrition and pest (disease and weed) management, together with soil and climatic conditions ensured the increase of crop productivity by 2.3 times—from 17.67 t/ha in 2000 to 41.08 t/ha in 2012. The area the sugar beets were harvested in the respective period decreased by 39.9 percent (from 747.0 thousand hectares in 2000 to 448.9 thousand hectares in 2012). The gross production of sugar beets increased by 39.7 percent (from 13,198.8 thousand tons in 2000 to 18,438.9 thousand tons in 2012). The increase of sugar beet gross production during this period was due to the increase of their productivity. Thus, the tendency of the area optimization under sugar beet crops along with the yield increase by the intensive use of land resources is observed in Ukraine.

As for wheat production, the yield of the crop increased by 1.4 times—from 1.98 t/ha in 2000 to 2.8 t/ha in 2012. The gross production of wheat increased by 54.6 percent (from 10,197.0 thousand tons the year 2000 15,762.6 thousand tons in 2012). The increase of wheat gross production is caused by the increase of productivity and that of the harvested area—by 9.1 percent (from 5161.6 thousand hectares in 2000 to 5629.7 thousand hectares in 2012). Accordingly, the corn yield for grain increased by 1.6 times—from 3.01 t/ha in 2000 to 4.79 t/ha in 2012. The gross production of corn increased by 5.4 times (from 3848.1 thousand tons in 2000 to 20,961.3 thousand tons in 2012). The increase of the corn gross production to a large extent is due to the increase of the harvested area by 3.4 times (from 1278.8 thousand ha in 2000 to 4371.9 thousand ha in 2012).

The market condition of agricultural crops remains unattractive for sugar beet growers (Table 22.2) [12–16]. The competition with other crops (wheat and corn) pushes the farmers to a situation wherein they have to reduce the sugar beet fields. Thus, the price of wheat increased from $166.5/t in 2011 to $194.6/t in 2012, or 16.9 percent, respectively, for corn—from $170.0/t to $190.3/t, or 12.0 percent, while the prices of sugar beet decreased from $65.2/t in 2011 to $53.8/t in 2012, or 17.5 percent. Within a year, the level of wheat production profitability declined from 17.6 to 11.8 percent, that of corn—from 38.6 to 19.8 percent, and sugar beet—from 36.5 to 15.7 percent, respectively.

TABLE 22.2   The economic efficiency of sugar beet, wheat, and corn production in Ukraine in the years 2008–2012 (agricultural companies)

Crops production indices	2008	2009	2010	2011	2012	%, 2012 in comparison with	
						2008	2011
Sugar beets							
Total cost of 1 t, US $	39.7	39.2	52.6	47.7	46.5	117.1	97.5
Average price of 1 t, US $	42.5	53.7	61.4	65.2	53.8	126.6	82.5
Level of profitability (%)	7.1	37.0	16.7	36.5	15.7	-	-
Wheat							

TABLE 22.2   *(Continued)*

Crops production indices	2008	2009	2010	2011	2012	%, 2012 in comparison with	
						2008	2011
Total cost of 1 t, US $	122.1	96.6	125.5	141.6	174.1	142.6	123.0
Average price of 1 t, US $	143.6	102.2	137.5	166.5	194.6	135.5	116.9
Level of profitability (%)	17.6	5.7	9.6	17.6	11.8	-	-
Corn							
Total cost of 1 t, US $	125.9	92.0	120.5	122.6	158.9	126.2	129.6
Average price of 1 t, US $	139.2	111.7	156.5	170.0	190.3	136.7	112.0
Level of profitability (%)	10.6	21.5	29.9	38.6	19.8	-	-

In the context of the world, during the lack of foodstuff, which is an urgent problem, the international community may most likely prohibit the bioethanol production using corn and wheat. The byproducts of sugar beet processing are not used directly for food, which is a relevant confirmation of the expediency to use it for the bioethanol production. In addition, there will be no need of reduction of the sugar beet areas but rather expansion, which in turn will create additional jobs in the sugar beet industry.

Foreign experience of sugar beet production and analysis of the current state of the domestic sugar beet production show that an important factor for improving the competitiveness of production in the investigated area is the rational distribution of sugar beet fields. By resolving this issue, the yield and sugar content of roots will be improved.

On the basis of the soil potential (fertility of the main soil types), the peculiar features of the climatic conditions, which are determined by the interaction of factors such as incoming solar radiation, atmospheric circulation, moisture supply, the researchers of the Institute of Bioenergy Crops and Sugar Beets of the NAAS of Ukraine defined a beetroot zone—which is the most favorable zone (as to its soil and climatic conditions) for sugar beet cultivation [5].

The most favorable growing area for this important agriculture crop (the zone of sufficient moisture, the rainfall/precipitation is over 550 mm per year), which allows to produce the sugar beet yields within 55–60 t/ha, are the western regions of Ukraine—Volyn, Ivano–Frankivsk, Lviv, Rivne, Ternopil, and Khmelnytskyi. In the less favorable zone (unstable moisture, the rainfall is 450–480 mm per year) includes Vinnytsia, Zhytomyr, Kyiv, Poltava, Sumy, Cherkasy, and Chernihiv regions, we can obtain guaranteed yield of beets—50–55 t/ha [5].

Still less favorable area of beet cultivation zone (the zone of low moisture, the rainfall is 430–480 mm per year) is in Kirovohrad and Chernivtsi regions, where we can obtain the yields of sugar beets at 45–50 t/ha.

The rest of the regions, including those where sugar beets are cultivated, are not favorable for sugar beet production because of their soil and climatic conditions.

In Ukraine, the developed sugar beet production is a universal basis for the production of bioethanol (Table 22.3).

The greatest output of bioethanol per unit area at the appropriate level of yield can be obtained from the sugar beets. However, in the processing of sugar beet into sugar, we get the molasses, and depending on its quality, the output of bioethanol from 1 t can be 0.222–0.237 t.

Considering the world experience of using sugar beets for bioethanol production as an alternative fuel, it would be appropriate to implement it at the sugar processing factories of Ukraine. The need to diversify a subcomplex of sugar beet production is determined not only by the country's high dependence on energy resource import, but also by the necessity to have additional facilities to process the excess production, taking into account the cyclical and risk nature of sugar beet production.

TABLE 22.3 The calculation of the output of bioethanol from various types of raw materials by the different yields

Raw	The output of bioethanol from 1 t of production (t)	The output of bioethanol in calculating per 1 ha depending on the yields of culture (t)	
		Yield	Output of bioethanol
Sugar beets (crude juice)	0.074–0.079	40.0	2.96–3.16
		50.0	3.70–3.95
		60.0	4.44–4.74
Molasses (processing of sugar beet into sugar)	0.222–0.237	1.56	0.35–0.37
		1.95	0.43–0.46
		2.34	0.52–0.55
Wheat	0.237–0.311	3.0	0.71–0.93
		4.0	0.95–1.24
		5.0	1.19–1.56
Corn	0.321–0.346	4.0	1.28–1.38
		5.0	1.61–1.73
		6.0	1,93–2.08

The calculation of the cost of bioethanol production from different bioraw materials shows that the most competitive bioethanol production is from molasses (Table 22.4).

TABLE 22.4 The competitiveness of bioethanol production depending on the bioraw in Ukraine in 2012

Type of bioraw	The average price of 1 t bioraw (USD)	The need of bioraw for the production of 1 t of bioethanol (t)	Prime cost of bioethanol (USD)	
			1 t	1 l
Sugar beets	53.8	12.65–13.49	1447.5	1.14
Molasses	81.5	4.22–4.50	934.7	0.74
Wheat	194.6	3.21–4.22	1313.1	1.04
Corn	190.3	2.89–3.11	1184.0	0.94

Note. The official exchange rate of the National Bank of Ukraine in 2012: $ 100 = UAH 799.1.

On the basis of the polynomial model, which describes the level of sugar beet productivity in Ukraine within the period of 1913–2012, the prediction of sugar beet yields (Figure 22.1), which envisages its increase by 14 percent, was made.

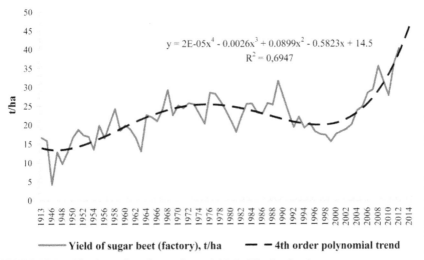

$$y = 2E\text{-}05x^4 - 0.0026x^3 + 0.0899x^2 - 0.5823x + 14.5$$
$$R^2 = 0,6947$$

——— **Yield of sugar beet (factory), t/ha**   — — **4th order polynomial trend**

FIGURE 22.1 The dynamics of sugar beet yields in Ukraine for the years.

The intensive consumption growth and condition of bioethanol world market enable to quickly build up the capacity of its production in Ukraine.

The European Union countries are interested in bioethanol import from our country. Every year the bioethanol market is growing by 3 percent in Europe, this is a significant potential for Ukraine to increase the export [3].

Thus, there are the prerequisites for biofuel application in Ukraine, in particular, by means of the diversification of the output production of sugar beet industry.

The competitiveness of bioethanol production using the output of sugar beet processing depends on several factors: the prices of sugar beets, the quality of the processed products of sweet roots, and the technology of its manufacture.

In Ukraine, the price of sweet roots affects the competitiveness of bioethanol production from the sugar beet processing products significantly.

Thus, when the price of roots was \$42.3/t, the cost of bioethanol production was \$1316.1/1 t (\$1.04 per 1 l). By raising the price of the sugar beet by 51.5 percent to \$64.1/t, the cost of bioethanol production was \$1566.1/1 t (\$1.24/1 l) or increased by 19 percent (Table 22.5).

TABLE 22.5 The calculation of the cost of bioethanol production from sugar beet (crude juice) depending on their prices in Ukraine in 2012

Index	Sugar beets price (USD/t)		
	Minimum	Medium	Maximum
	42.3	**53.8**	**64.1**
Raw material expenses, $/t	724.8	856.1	974.8
Processing of raw material, $/t	591.3	591.3	591.3
Production cost of bioethanol, $/t	1316.1	1447.5	1566.1
$/l	1.04	1.14	1.24

Note. The official exchange rate of the National Bank of Ukraine in 2012: \$100 = UAH 799.1.

The similar dependence is observed when bioethanol is produced from the molasses in Ukraine (Table 22.6).

TABLE 22.6 The calculation of the cost of bioethanol production from molasses depending on its price in Ukraine in 2012

Index	Molasses price (USD/t)		
	Minimum	Medium	Maximum
	71.1	**81.5**	**90.9**
Raw material expenses, $/t	299.6	343.4	383.0

TABLE 22.6 *(Continued)*

Index	Molasses price (USD/t)		
	Minimum	**Medium**	**Maximum**
	71.1	**81.5**	**90.9**
Processing of raw materials, $/t	591.3	591.3	591.3
Production cost of bioethanol, $/t	890.9	934.7	974.3
$/l	0.70	0.74	0.76

Note. The official exchange rate of the National Bank of Ukraine in 2012: $100 = UAH 799.1

When the price of molasses increases by 27.8 percent—from $71.1 to 90.9/t, the cost of bioethanol production will increase by 9.4 percent—from $890.9 to 974.3 per 1 t, or $0.7–0.76 per 1 l.

These research studies have shown that provided the molasses is not a commodity and in turn it has no price, and sugar refineries process it into bioethanol, then its cost of production will decrease significantly.

In addition, the combined production of sugar for the consumer market and bioethanol from the molasses enhances the competitiveness of the sugar beet industry enterprises considerably. Thus, when the price of sweet roots is $53.8/t, the factory can produce sugar by the price of $698.5/t and bioethanol from the molasses—$934.7/t; however, if it produces only sugar or bioethanol from sugar beet, then the gross income will significantly reduce. The similar correlation is observed when the price of sugar beet is different (Table 22.7).

TABLE 22.7 The cost of sugar, molasses, bioethanol production depending on the price of sugar beet in Ukraine in 2012

Produce	Sugar beet price (USD/t)			Coefficients of ratios
	Minimum	**Medium**	**Maximum**	
Sugar beets	42.3	53.8	64.1	1.0
The cost of sugar, molasses, and bioethanol production (USD/t)				
Sugar	608.7	698.5	778.6	12.1–14.4
Molasses	71.1	81.5	90.9	1.4–1.7
Bioethanol from molasses	890.9	934.7	974.3	15.2–21.1
The cost of bioethanol production from sugar beet (crude juice) (USD/t)				
Bioethanol from beets	1316.1	1447.5	1566.1	24.4–31.1

Note. The official exchange rate of the National Bank of Ukraine in 2012: $100 = UAH 799.1.

The experience of the foreign countries and price situation of sugar beet in Ukraine indicate that to develop the technologies, which use cheaper and economically better raw material for bioethanol production, is an important task for sugar beet production as the use of sugar beet and the products of their processing for the manufacture of other commodities, besides sugar, makes them more competitive in comparison with other crops.

The sugar beet production should be concentrated at large well-developed farms that are able to implement intensive technologies with the use of modern agricultural machinery, fertilizers, integrated crop, and soil protection, improve sugar beet yield and quality, and reduce their costs as important factors of sugar quality improvement according to the requirements of the consumer sugar, which exist in EU countries, and ensuring the competitiveness of sugar in the domestic and foreign markets.

The quality and the type of raw material have the effect on the increase of the competitiveness of bioethanol production.

The raw material for ethanol production, besides molasses and sugar beets, can be the byproducts of sugar beets with high sugar content, in particular, the green black-strap molasses and syrup. Processing of 1 t of green blackstrap molasses—394 l or 311 kg and 1 t of syrup—375 l (297 kg) gives the largest output of bioethanol. The output of bioethanol from molasses is 1 t—300 l (237 kg), 1 ton of sugar beet—100 l (79 kg).

The reduction of the bioethanol cost and the increase of its competitiveness depend greatly on the production technology. Bioethanol technology consists of two phases: the production of raw ethanol and its further dehydration. Azeotropic distillation, adsorption on molecular sieves, and evaporation through the membrane are used for ethanol dehydration of Shiyan [8].

The cost of bioethanol production from sugar beet (crude juice), green molasses, syrup and molasses, depending on the technology used, was calculated by us (Table 22.8).

TABLE 22.8   The competitiveness of bioethanol production from the products of sugar beet processing with the use of different technologies in 2012

| Bioraw | Production cost of bioethanol using different technologies ($) | | | | | |
| | Azeotropic distillation | | Adsorption on molecular sieves | | Evaporation through the membrane | |
	1 t	1 l	1 t	1 l	1 t	1 l
Sugar beets	1447.5	1.14	1409.9	1.11	1378.2	1.09
Green blackstrap molasses	1194.8	0.94	1157.3	0.91	1125.6	0.89
Syrup	1646.4	1.30	1608.8	1.26	1577.1	1.24
Molasses	934.7	0.74	897.2	0.71	865.5	0.69

Note. The official exchange rate of the National Bank of Ukraine in 2012: $100 = UAH 799.1

Bioethanol with the lowest cost is obtained by the evaporation through the membrane independent of the type of raw material for processing. Thus, when the bioethanol production from the sugar beet (crude juice) by the azeotropic distillation was used, the production cost of 1 ton of bioethanol was USD 1447.5 ($1.14 per 1 l), and when the evaporation through the membrane was used—$1378.2 ($1.09 per 1 l) or decreased by 4.8 percent.

The analogical dependence occurred when the bioethanol production from green blackstrap molasses, syrup, and molasses took place. The lowest production cost of bioethanol from molasses was obtained while using all three processing technologies. Thus, the production of bioethanol from the molasses by azeotropic distillation the production cost of 1 ton of bioethanol was $934.7 ($0.74 per 1 l), and it decreased by 7.4 percent–$865.5/t ($0.69 per 1 l) when the evaporation through the membrane was used.

The calculations of the bioethanol production cost from sugar beet byproducts show that the competitiveness ensures the bioethanol production from molasses using the evaporation through the membrane.

It should be noted that the dehydration of the ethanol by the azeotropic rectification requires considerable operational and energy expenses. Ethanol dehydration technologies using the adsorption on molecular sieves and evaporation through the membrane are less power-consuming ones. However, the ethanol dehydration by the evaporation through the membrane requires significant capital investment and the smooth/uninterrupted operation of the factory.

According to the data of Ukraine's State Statistics, the average annual consumption of gasoline in the country varies within 4.7 million tons within years, including agriculture—0.2 million tons (Table 22.9) [17–21].

TABLE 22.9 Fuel consumption and calculation of bioethanol needs in Ukraine

| Year | Consumed gasoline | | The needs of bioethanol for mixtures with gasoline | | | |
| | | | 6% | | 10% | |
	Total (thousand tons)	Including agriculture (thousand tons)	Total (thousand tons)	Including the agriculture, thousand tons	Total (thousand tons)	Including the agriculture (thousand tons)
2007	4821.8	299.0	289.3	17.9	482.2	29.9
2008	5061.1	288.5	303.7	17.3	506.1	28.9
2009	4696.1	216.0	281.8	13.0	469.6	21.6
2010	4632.7	231.6	278.0	13.9	463.3	23.2
2011	4401.3	224.5	264.1	13.5	440.1	22.5

With the small deviations, these amounts of gasoline are forecasted for future. There are no significant disruptions in providing agricultural enterprises with this type of fuel. However, the price of gasoline is growing by 7 percent almost every year. Specific annual increase of fuel prices occurs in April and September, which coincides with the main field processes in agriculture—sowing and harvesting time. Owing to low solvency of agricultural enterprises, it may have a negative effect on proper timing of production processes. And the price of oil increases worldwide every year. Therefore, the solution consists in production and use of alternative fuel types, first of all for agriculture.

The growth of sugar beet production by 2015 and 2020 will facilitate the implementation of the Law of Ukraine "On Alternative Types of Fuels," in the part of bioethanol addition into gasoline, as well as reduce the dependence of agricultural producers on the market prices of fuel and its import (Table 22.10).

TABLE 22.10   The estimated calculation of bioethanol production based on the productivity of sugar beet in Ukraine

	Forecast			
Index	2015		2020	
	Optimistic	Pessimistic	Optimistic	Pessimistic
Gasoline consumption, total, thousand tons	4,723		4,723	
Including agriculture, thousand tons	252		252	
Share of bioethanol in gasoline mixtures,%	6		10	
Required bioethanol amount, thousand tons	283		472	
Including agriculture, thousand tons	15.1		25.2	
Sugar beet production, thousand tons	26,000	18,200	35,000	25,000
Processing of sugar beets into sugar, thousand tons	15,090		14,775	
Bioethanol, thousand tons	10,000	2,473	19,000	9,350
Molasses output, thousand tons	604		473	
Bioethanol production from molasses, thousand tons	143		112	
Bioethanol production from beets, thousand tons	790	195	1,501	739
Bioethanol production, total, thousand tons	933	338	1,613	851

By the foretold forecast by the year of 2020, the significant growth of agricultural production in Ukraine is creating favorable conditions for the renewable energy development, in particular for the increase of biofuel production [22]. Also, the steady fuel price increase promotes fuel production and use in Ukraine.

In order to ensure the predicted volumes of bioethanol production in Ukraine, it is necessary to stimulate the producers to do the renewal of fixed assets for both sugar beet and sugar and bioethanol production. The best option is the construction of the combined workshops at sugar and ethanol factories, which will produce bioethanol from sugar beet byproducts in the season of sugar beet harvesting, and in the off-season—from the waste of grain of headed crops or corn.

The solution for these issues requires the development of a national comprehensive program for sugar industry enhancement in Ukraine, which will take into account world tendencies as to the use of sugar beet produce for bioethanol manufacture as an alternative fuel.

22.4 CONCLUSIONS

Sugar beet high yields are most likely produced by the application of intensive technologies with the use of all agrobiological cultivation practices together with their high-quality performance in optimal terms. In Ukraine, sugar beet production has to be concentrated in the most favorable areas for sugar beet cultivation, where soil and climatic conditions guarantee high indicators of root productivity and quality.

The bioethanol production and use, in particular from the byproducts of sugar beet processing, will allow to establish the production of environmentally clean alternative fuel, create new jobs, increase the profits of the enterprises, reduce the dependence of Ukraine on imported fuel, and develop new markets for the products of sugar beet manufacture. Producing bioethanol from molasses can meet the demand for this product at a lower price and improve the competitiveness of the enterprises of sugar beet industry as well. The solution for these issues will increase the competitiveness of the produce manufactured by businesses of AIC (agroindustrial complex) of Ukraine both in domestic and foreign markets.

ACKNOWLEDGEMENTS

The author expresses his gratitude for the help in preparing materials to National Association of Sugar Producers of Ukraine "Ukrtsukor" and personally to Mykola Yarchuk, Petro Borysiuk, and Mykola Kalinichenko. Special thanks goes to Vitaliy Sosnitskyi for consultation and help on the issues of innovative technologies of ethanol industry. Sincere words of gratitude are for his scientific advisor, Prof. Maria Kodenska.

KEYWORDS

- **Bioethanol**
- **Competitiveness**
- **Efficiency**
- **Molasses**
- **Production cost**
- **Sugar**
- **Sugar beets**
- **Sugar beet production**

REFERENCES

1. Varchenko, O. M.; The World and National Experiences of the Sugar Market Regulation: Monograph. Bila Tserkva; **2009,** 334 p. (in Ukrainian).
2. Zayets, O. S.; The Sugar Market: Issues, Trends, and Practices. Kyiv: "Naukova Dumka"; **1998,** 365 p. (in Ukrainian).
3. Kaletnik, G. M.; The perspectives of bioethanol production in Ukraine. *Agric. Equip. Machinery.* 2009, *2,* 50–55. (in Ukrainian).
4. Kodenska, M. Yu.; The Justification of the Need of the Investment Projects Development in Sector of Bioethanol Production Development Based on Sugar Beet Production. AhroInKom; **2011,** 4–6 [Electronic resource]. - Mode of access: http://www.nbuv.gov.ua/portal/Chem_Biol / Agroin/2011_4-6/KODENSKA.pdf, free. - From the screen. (in Ukrainian).
5. Royik, M. V.; Beets. Kyiv: "XXI vik—RIA TRUD-KYIV"; **2001,** 320 p. (in Ukrainian).
6. Sugar beet production in Ukraine: problems of revival, perspectives of development Monograph. Sabluk, P. T.; Kodenska, M. Y.; Vlasov, V. I.; et al., Ed. Sabluk, P. T.; Kodenskoyi, M.; Kyiv: "IAE NNC"; **2007,** 390 p. (in Ukrainian).
7. Spychak, A. M.; The Economic problems of biofuel production and food security of Ukraine. *Economy APK.* **2009,** *8,* 11–19. (in Ukrainian).
8. Shiyan, P. L.; Sosnitsky, V. V.; and Oliynichuk, S. T.; The innovative technologies of alcohol industry. Theory and Practice: Monograph. Kyiv: "Askaniya"; **2009,** 424 p. (in Ukrainian).
9. Directive 2009/28/EC of the european parliament and of the council of 23 April 2009 on the promotion of the use of energy from renewable sources and amending and subsequently repealing Directives 2001/77/EC and 2003/30/EC [Electronic resource]. - Mode of access: http://eurlex. europa.eu/LexUriServ/LexUriServ.do?uri=OJ:L:2009:140:0016:0062:EN:PDF, free. - From the screen.
10. The Law of Ukraine "About Alternative Types of Fuel," from January 14th, 2000 No 1391–XIV: as of the 21.07.2012. The Verkhovna Rada of Ukraine. Off. Publ. Kyiv: Parliamentary Publishing House; **2000,** *12,* 94 (The Library of Official Publications) (in Ukrainian).
11. Shiyan, P. L.; The Plant Growing of Ukraine: Statistical Digest for 2012. Ed. Vlasenko, N. S.; Kyiv: State Statistics Committee of Ukraine; **2013,** 110 p. (in Ukrainian).
12. Shiyan, P. L.; The Basic Economic Indexes of Agricultural Production in Agricultural Enterprises: The Statistical Bulletin for 2008. Kyiv: State Statistics Committee of Ukraine; **2009,** 76 p. (in Ukrainian).
13. Shiyan, P. L.; The Basic Economic Indexes of Agricultural Production in Agricultural Enterprises: The Statistical Bulletin for 2009. Kyiv: State Statistics Committee of Ukraine; **2010,** 80 p. (in Ukrainian).

14. Shiyan, P. L.; The Basic Economic Indexes of Agricultural Production in Agricultural Enterprises: The Statistical Bulletin for 2010. Kyiv: State Statistics Committee of Ukraine; **2011, 88** p. (in Ukrainian).
15. Shiyan, P. L.; The Basic Economic Indexes of Agricultural Production in Agricultural Enterprises: The Statistical Bulletin for 2011. Kyiv: State Statistics Committee of Ukraine. **2012, 88** p. (in Ukrainian).
16. Shiyan, P. L.; The Basic Economic Indexes of Agricultural Production in Agricultural Enterprises: The Statistical Bulletin for 2012. Kyiv: State Statistics Committee of Ukraine; **2013, 88** p. (in Ukrainian).
17. Shiyan, P. L.; The Statistical Yearbook of Ukraine for the 2007. Ed. Osaulenko, O. G.; Kyiv: State Statistics Committee of Ukraine; **2008,** 571 p. (in Ukrainian).
18. Shiyan, P. L.; The Statistical Yearbook of Ukraine for the 2008. Ed. Osaulenko, O. G.; Kyiv: State Statistics Committee of Ukraine; **2009,** 566 p. (in Ukrainian).
19. Shiyan, P. L.; The Statistical Yearbook of Ukraine for the 2009. Ed. Osaulenko, O. G.; Kyiv: State Statistics Committee of Ukraine; **2010,** 566 p. (in Ukrainian).
20. The Statistical Yearbook of Ukraine for the 2010. Ed. Osaulenko, O. G.; Kyiv: State Statistics Committee of Ukraine; **2011,** 559 p. (in Ukrainian).
21. Shiyan, P. L.; The Statistical Yearbook of Ukraine for the 2011. Ed. Osaulenko, Kyiv: State Statistics Committee of Ukraine; **2012,** 558 p. (in Ukrainian).
22. The Strategic Directions of Agriculture Development for the Period Till 2020 in Ukraine. Ed. Lupenko, Y. O.; Mesel–Veselyaka, V. J.; 2nd Publication. Revis. and Enlar. Kyiv: NSC IAE; **2012,** 218 p. (in Ukrainian).

PART IV
A ROLE OF CHEMICAL MUTAGENESIS IN INDUCTION OF BIODIVERSITY OF AGRICULTURAL PLANTS

CHEMICAL MUTAGENESIS AS A METHOD OF BASELINE GENETIC DIVERSITY ENLARGEMENT IN PLANT BREEDING

STANISLAV P. VASILKIVSKIY and ANATOLIY Y. YURCHENKO

CONTENTS

23.1 INTRODUCTION

Biological potential of a field depends totally on the variety of genetic system, which is manifested by the degree of effective use of ecological potential, which is created by humans with technical equipments; hence, rentable technical and soil-reclamation operations depend on it.

Vavilov [1] defined breeding as an evolution directed by human power. Breeding is based on the selection of plants with the changes caused by changes in their genotype. Understanding of the variability is of fundamental importance in studying the baseline, which is the most important part of breeding.

Wide use of modern methods of studying genetic effect enables us to penetrate into the most important processes, which make the base of heredity, to discover its regularity, as well as to solve one of the main tasks of breeding—creating the genetic diversity of baseline [2].

Gene and chromosome mutations are the main source of all heredity variabilities and the materials for evolution and breeding [3]. According to the law of homologous series proposed by M.I. Vavilov, the world collection of breeders can be reproduced by means of mutation [4].

Mutations are heredity change, making genetic base of variability and the raw material for selection. As Vavilov affirms [5], mutations are significant in winter wheat forms created by natural and artificial hybridization.

Mutation variability underlies any baseline for breeding, as the initial primary changeability arises only as a result of mutation. Breeding and artificial selection are based on the processes of spontaneous changeability of genes and chromosomes. Spontaneous mutational changeability is important in evolution of an organic world, although it arises with low frequency [6–8].

Primitive breeding, based on the selection of spontaneous mutations and hybrids for thousands of years, have created valuable forms of crops hard to be improved, even with the use of modern breeding methods.

Mutations are displayed in changes of plant organs (root, steam, leaf, inflorescence, flower, fruit, and seeds) and their study gives rich material for taxonomy. Experimental mutagenesis use enables to disclose all the possibilities of species in forms on the basis of received polymorphism and to create rich collections of genetically diversity of some plants species on the basis of the received mutations.

Mutation process is a "supplier" of new elementary material; it supports genetic heterogeneity of natural populations. However, plant breeding could not be based on using only spontaneous mutations as their frequency is quite low. The induced mutagenesis is a powerful method of solving multiple theoretical and practical tasks of genetics and breeding.

Methods of induced mutagenesis complement all the other parts of initial material studied, as they enable to involve the variegated material in the form of gene and chromosome mutations into the breeding, although the mutations are just a baseline as they can give rise to new varieties only after they have undergone strict selection, in some cases even with hybridization [9].

Experiments obtaining mutations and their applications in plants breeding have attracted attention only after discovering high mutagen activity of ionizing radiation in mushrooms [9, 10] and higher plants [11].

Rapoport [12–15] and Auerbah [16] discovered a lot of new highly active mutagens. In his further scientific works, starting from 1957 at the Institute of Chemical Physics of the USSR Academy of Sciences, I.A. Papoport discovered another large amount of highly active substances—supermutagens, which cause hundreds and thousand times more mutations compared with nonorganic chemical compounds. Williams [17] accentuates the importance of mutations for breeding, stating that plant breeding is based on the definition and use of mutations that strengthen their characteristics in the desired way.

Nowadays, hundreds of chemical compounds with mutagen properties are known. Soyfer [18], while studying chemical mutagens action mechanism, discovered alkylate compound reaction with both nitrogen base and DNA phosphate groups. These reactions can finally result in destroying purine base, which may cause spot mutation and DNA sugarphosphatic skeleton breakage, which causes chromosome aberration.

Chemical mutagens are strong factors inducing changes in genetic material.

Primary damages of DNA molecules with chemical mutagens include disturbance in molecule structure (acridine inclusion and phosphorus group alkylation), transformation of DNA normal base into atypical one (either because of their chemical modification or because of their change in the process of replication on the structural analogue), and destruction of bases. These damages can be repaired, saved, or transformed into new changes of molecules and exhibited through change of their characteristics. Specificity of these changes can exhibit in the initial physiological and genetic effects: early or distant death of plants; growth and development slowdown or stimulation; and frequency of chromosome reformation.

Induced mutations make a wide spectrum of forms by changing any agricultural and farm-valuable features of crops. That is why quite an easy way to work out the weak points in plant breeding of any crop seemed possible. That is, a perspective of improvement of single characteristic with simultaneous saving, the complex of other useful practical characteristics in breeding varieties was considered. Yet, it turned out that working out the weak points, that is to improve one or a few characteristics possible is not always possible. Mutation changes through mutation effect and other interaction of genes often envelop in the entire organism. As a result, undesirable agricultural or breeding features may arise along with the newly developed ones.

Research aims. The research aims at detecting cytogenetic activity of chemical mutagens depending on the genotype variety and proving the possibility of selection and identification of winter wheat lines of mutagen origin, selected for morphological characteristics in generations of genetically unstable mutants.

23.2 MATERIALS AND METHODOLOGY

Investigated cytogenetic and mutagen action of eight alkylate DNA compounds: N-nitozo-N-methyl urea (NMU), N-nitrozo-N-ethyl urea (NEU), N-nitrozodymethyl urea (NDMU), dimethyl sulfate (DMS), diethyl sulfate (DES), nitrozomethylbiuret (NMB), 1,4-bisdiazoacetylbutane (DAB), and ethylene imine (EI). The mutagens synthesized at the Department of Chemical Mutagenesis of the Institute of Chemical Physics (Moscow) were kindly granted for our research. Mutagen influence of these compounds was investigated in Ukrainian and foreign varieties and in the variety mutants M_3–M_7 induced by us. During the research period (1981–2012), we studied 50 variety samples that underwent the influence of mutagens.

Dry seeds of the variety samples were soaked in room temperature for 4–18 h in water solutions with mutagen concentration values of 0.05, 0.025, and 0.0125 percent. After soaking, the seeds were washed with water for an hour, air-dried, and 500 seeds of each variant were sowed in the field and 100 seeds—in Petri dish for cytological analysis. The soaked seeds were the control.

Cytogenetic activity of mutagens was studied by frequency of cytogenetic aberrations in meristem cells of initial roots of seeds and soaked in mutagen solutions. Then they were fixed in solution of Karnua changed fixer on temporary acetocarmine substances in late anaphase and early telophase. Minimum 600 cells were examined for each variant.

M_1 plants were observed from their germination to maturity. Changed forms account in M_1 plants were accounted on their height and the amount of changed families was counted in M_2.

We chose plants with morphological changes, made individual biometrical analysis, and the seeds were sowed for the heredity checking in M_2 and further generations. The rest of M_1 plants were gathered, individual analysis was made, and spike sown (main spike) that guaranteed avoiding mechanical infestation with other forms.

Seeds from individually chosen plants in M_1–M_3 plants with morphological changes were sowed in baseline nursery for their further examination compared with initial varieties.

In our research study with constant mutagens, we used individual selection for breeding and agricultural valuable characteristics; in nonstable, the continuous selection by pedigree method in combination with mass selection of plus-variants in 39 generations after selection of mutant plant was used.

Identification of genotypes by electrophoresis spectrum of gliadins was conducted by the results of the analysis of 100 kernels of each line. Electrophoresis of gliadins in polyacrilamide gel (PAAG) was performed in "Molecular markers" laboratory of Bila Tserkva National Agrarian University.

Biometrical analysis was made by well-known methods on an average sample of 20–30 plants. Statistic analysis was performed by methods of descriptive statistics, correlation, and variance analysis with "Statistic" v.6.0 program.

23.3　RESULTS AND DISCUSSION

Strong chemical mutagens are characterized with their different influence on plants cells. Many of them have cytotoxic impact on all levels of cell organization. Mechanism of cytotoxic effect shows a wide range of interaction of chemical agents that can enter into reaction with both gene material and cytoplasmic enzymes. So, toxic influence of mutagens can be a reason for structural damage of chromosomes or for inactivation of essential enzymes.

Molecules of mutagen while penetrating through cell membrane enter the metabolic balance of organism and condition different disorders in the in-cell balance. As chemical mutagens influence on viability is revealed more intensive on the first stage of growth and development of M_1 plants, we also studied intensity of roots growth initially in the period of seed germination in laboratory conditions besides well-known criteria of plant sensitivity to mutagens (field germination of seeds, survival, and sterility of plants).

By the treatment by mutagen, water penetrates the seed tissue first, with mutagen molecules to follow. Individual field of mutagen molecules consists of main physical and chemical constants and expects not only specific interaction with gene material but also with structural elements of nucleo- and cytoplasm for each mutagen as well [19].

Mutations arise as a result of mutagen interaction with cell genetic material. On the way to this contact, mutagens interfere the metabolism balance and can cause different abnormalities of in-cell balance and injury of genome part. That is why the first processes that take place directly after seeds soaking in mutagen solution cause injury on the cell level and violation in chromosome structure.

Chromosome aberrations are important in genetic variation, monitoring, and evolution and are one of the main characteristics of mutagen genetic activity. We studied the influence of six mutagens on five varieties by frequency of chromosome aberrations. Mutagens break a chromosome apparatus and normal behavior of mitosis in initial roots cells. The violation frequency depends on the concentration of mutagen. Direct relation was detected in all varieties and mutagens: chromosome aberration frequency increases with the increase in mutagen concentration For example, NEU 0.0125, 0.025, of 0.5 percent concentration conditioned from 3.7 percent of chromosome aberrations in Illichivka variety to 6.4 percent—in Krasnodarska 46 one, from 8.2 percent in Illichivka variety to 10.9 percent in Polisska 70 one and from 14.1 percent in Myronivska 25 variety to 18.7 percent in Illichivka, respectively. Ethylene imine in these varieties conditioned more violations in chromosome structure compared with other varieties (Table 23.1).

Mutagens also differ in their genetic activity. Thus, 0.05 percent NMB conditioned from 18.2 ± 1.5 to 31.8 ± 1.8 percent in five varieties, while DES of the same concentration—only from 12.3 ± 1.3 to 16.7 ± 1.5 percent of chromosome aberrations. NMB and EI revealed the highest activity by their influence on mitotic cell apparatus of all the studied mutagens.

Mutagen activity is revealed uneven in different genotypes (Table 23.1). In Polisska 70, Krasnodarska 46, and Illichivka varieties, the highest percentage of chromosome aberrations was conditioned with NMB 0.05 percent concentration while in the seed of these varieties treated with NMB and EI 0.0125 percent EI concentration solution, it conditioned violation of chromosome structure with higher frequency than with NMS.

TABLE 23.1 Chromosome aberration frequency in winter wheat initial roots meristem cells treated with mutagens

Concentration		Poliska 70	Krasno-darska 46	Illichivka	Myronivs-ka 61	Kiyanka
				Variety		
	H_2O–control	0.38 ± 0.3	0.15 ± 0.15	0.17 ± 0.2	0.44 ± 0.8	0.8 ± 0.4
HMU	0.05	29.9 ± 1.5	31.8 ± 1.8	27.0 ± 1.7	19.7 ± 1.6	18.2 ± 1.5
	0.025	23.1 ± 1.6	16.8 ± 1.6	16.4 ± 1.4	14.9 ± 1.4	8.6 ± 1.1
	0.0125	6.5 ± 1.0	9.6 ± 1.1	11.1 ± 1.1	8.4 ± 1.2	6.7 ± 1.0
NEU	0.05	18.5 ± 1.5	16.5 ± 1.5	18.7 ± 1.6	14.1 ± 1.4	15.3 ± 1.3
	0.025	10.9 ± 1.2	9.2 ± 1.3	8.2 ± 1.1	12.0 ± 1.3	8.5 ± 1.1
	0.0125	5.2 ± 0.9	6.4 ± 1.0	3.7 ± 0.8	4.6 ± 1.2	5.4 ± 0.9
NDMU	0.05	16.2 ± 1.4	17.8 ± 1.4	10.3 ± 1.2	13.7 ± 1.3	12.2 ± 1.2
	0.025	10.5 ± 1.2	3.7 ± 0.8	8.3 ± 1.1	6.1 ± 0.9	9.5 ± 1.2
	0.0125	4.1 ± 0.8	2.2 ± 0.6	3.1 ± 0.7	4.7 ± 0.9	4.7 ± 0.9
NMB	0.05	22.6 ± 1.6	24.2 ± 1.6	14.0 ± 1.4	10.1 ± 1.2	8.0 ± 1.0
	0.025	11.8 ± 1.3	11.4 ± 1.2	8.6 ± 1.2	7.3 ± 1.1	5.2 ± 0.9
	0.0125	6.2 ± 1.1	7.8 ± 1.1	3.6 ± 0.7	5.2 ± 0.9	2.9 ± 0.7
DES	0.05	16.7 ± 1.5	15.4 ± 1.3	15.9 ± 1.5	15.4 ± 1.3	12.03 ± 1.3
	0.025	5.6 ± 1.0	6.7 ± 1.0	9.5 ± 1.1	8.7 ± 1.9	7.4 ± 1.1
	0.0125	3.9 ± 0.8	4.9 ± 0.9	4.2 ± 0.8	6.2 ± 1.0	5.9 ± 0.8
EI	0.05	28.6 ± 1.4	30.6 ± 1.7	26.2 ± 1.6	23.2 ± 1.7	21.6 ± 1.5
	0.025	19.3 ± 1.2	21.8 ± 1.5	18.5 ± 1.3	15.3 ± 1.5	13.6 ± 1.4
	0.0125	12.4 ± 1.1	14.5 ± 1.3	12.4 ± 1.3	6.2 ± 1.0	8.9 ± 1.1

The highest activity by chromosome aberration frequency had EI 0.025 and 0.05 percent concentration in Myronivska 25 variety, and under different concentrations in Kiyanka. The lowest number of violations of chromosome structure caused NMB under all concentrations in Kiyanka variety, and in Poliska 70 and Krasnodarska 46 varieties, it was the third by this feature.

Our data match with the those of the other scientists in some way. By comparing the aberration activity of *N*-nitrozo-*N*-alkilurea (NMU, NDMU, NEU, NDEU, and NMB) on spring wheat varieties (Leningradka, Diamant), they had higher activity of NMU in all the concentrations. The rest mutagens, especially in NMB, had aberration frequency increase reliably only in tough doses.

Analysis of correlation between chromosome aberration frequency and field germination in our experiments revealed the inverse relationship. Correlation coefficients in different varieties ranging from $r = -0.473$ in Poliska 70 variety to $r = -0.704$ in Mironivska 25 prove the correlation dependence of field germination and chromosome aberration frequency in initial root meristem of mutagen-treated seeds.

Use of chemical compounds that cause mutations gave the breeders a powerful method of growth of diversity increase and enabled them to create valuable forms of crops. Collections of valuable mutants of barley [20, 21], buckwheat [22, 23], maize [24], wheat [25–27], and other valuable crops were created in a rather short time. Creating and saving a rich genetic fund of mutants is an urgent task as it is the material base for theoretical research development and practical breeding.

Identification of varieties, hybrids, and genotypes of cultural plants is one of the most important problems of agriculture. Varieties were initially differentiated only by most visible morphological characteristics. As scientific breeding methods arose, methods of genotypes, fast identification and variety purity identification of sowing materials have become significant in the process of new breeding material and varieties creation.

Thorough genetic studying of induced mutations is a guarantee of their successful use as induced mutants are new forms with changed genetically systems which have disturbed linkage between an organism and the environment and correlation between some features, created in the initial varieties by natural and artificial selection in breeding.

Selection of mutants in mutational breeding is based on morphological characteristics solely. However, the leveling of mutagen origin lines by the phenotype does not reflect its internal genetic structure. Search for molecular markers for identification of mutants is an urgent issue. Nowadays, use of molecular genetic markers for studying the tendencies of forming adopted gene complex in the process of selection, detecting the relationship between allele variant cluster of genes of reserve protein with loci, controlling the level of expression of quantitative characteristics, [28] for identification of genotypes and estimating the varietal purity [29] are introduced. Yet, these research studies were made on breeding varieties, intervarietal and interspecific hybrids, and lines of hybrid origin. The available literature does not submit a lot of publications on the usage of molecular markers for identification of breeding material of mutagen origin.

Among the biochemical gene markers, the ones controlling the reserve proteins have been studied better than the others. Reserve proteins of winter wheat—gliadins and high-molecular glutenins—have played a significant role in genetically breeding research studies of this culture lately. They have the only of the most informative genetic markers among the well-known ones [30]. Proteins are primary products of genetic systems, and protein electrophoretic activity is finally determined by the sequence of gene nucleotide structure. That is why proteins are markers of both single structural genes and their associations. Allele variants of proteins are usually inherited codominantly; their content is not predetermined with growing conditions and is detected comparatively easily by electrophoresis method [31].

We started the research studies on identification by electrophoretic spectrum of gliadins in 2001 and they have been carried over until now. Collection of mutants used as a baseline was set up in the previous years at the Department of Genetics, Breeding and Seed production of Bila Tserkva National Agrarian University. In 1998, S. Vasilkivskiy by the method of chemical mutagenesis bred from Czech Republic Roxana variety diazoacetylbutane 0.025 percent concentration induced and selected 432/5-dwarf mutant, which was not stable. The lines, which gave rise to new ones, were bred in its generations.

The 432/5 mutant belongs to real dwarf species by its morphological characteristics. The result of electrophoresis (Figure 23.1) results in conclusion about heterogeneity by number, intensity, and electrophoretic activity of compounds.

The M 432/5 contains no new components, which are missing in Roxana variety.

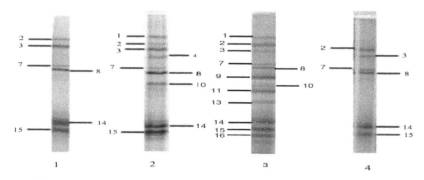

FIGURE 23.1   Electrophoretic spectra of winter wheat gliadins for 432/5 mutant.

In their practical work with mutants, research studies reject the forms, which in their fourth–fifth generations give splitting with deviations in their quantity correlations from the theoretically expected. We managed to work out a way of using new source of genetic changeability and to offer the method of working with nonstable mutants attributed to years of research at the Department of Genetics, Breeding and Seed Production of Bila Tserkva National Agrarian University [32, 33, 34].

In 1999–2005 in posterity of M 432/5 dwarf erytrospermum form, plants were individually chosen, different in their anatomy morphological and agricultural features, one of them gave rise to a new 710 erytrospermum line, another—for 712 erytrospermum line.

Presence of the same components in these lines points on their affinity, affinity with 432/5 mutant and with Roxana initial variety. Absence of some components proves their difference in both genetic and morphological levels.

Electrophoregrams of 710 erytrospermum and 712 erytrospermum samples also do not contain any new components, which are not identified in Roxana variety and M 432/5 mutant (Figure 23.2).

Lines 710 and 712 belong to the same form. However, components 4 and 5, present in line 712, is missed in line 710, which also proves the genetic difference between them. Electrophoresis of 712 erytrospermum variety sample showed that this sample is heterogeneous, and two biotypes were detected in it. Presence of identically located common components of electrophoregrams, in Roxana initial variety and lines 708–712, which were chosen from nonstable 432/5 dwarf mutant, points out their genetically similarity.

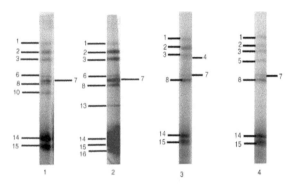

FIGURE 23.2   Electrophoretic spectra of gliadins; M 710 erytrospermum—1, 2; M 712 erytrospermum—3, 4.

Electrophoresis enables us to see practically all polymorphisms of proteins and identify new associations of genes. In our research studies of electrophoresis of gliadins, we detected heterogeneity and revealed the affinity with the initial variety of line chosen by the phenotype criteria markers.

Gliadine electrophoregrams of number 712 erytrospermum showed that this sample is heterogenic; two biotypes were detected by the components of gliadins (Figure 23.2, lines 3, 4). Lines 710 and 712 belong to the same form—erytrospermum. However, line 710 does not have components 4,5 in line 712, which also prove differences in their farm-valuable features. There are 19 movable components in line 712 which is present in both biotypes. Components 1, 2, 3, 6, 7, 8, 14, and 15 are present in Roxana variety and in line 710. Components 1, 2, 3, 7, 8, 14, and 15 were common for line 712 and Roxana variety. Presence of placement common components of gliadins 1, 2, 3, 7, 14, and 15 in Roxana initial variety and lines 708, 710, and 712 chosen from nonstable 432/5 dwarf mutant shows their genetic affinity with Roxana variety.

The research results prove that electrophoresis method of gliadins in polyacrylamide gel can be used for identification of winter wheat mutant forms.

23.4   CONCLUSIONS

Cytogenetic activity of mutagens in M_0 by their chromosome aberration frequency in winter wheat correlates with their concentration and exposition of seed soaking. The highest activity on the influence on cells mitotic apparatus reveal NMU and EI, and the least number of chromosome structure violation was caused by DEU and NMB. The method of gliadine electrophoresis in polyacrylamide gel can be used in genetic identification of mutant origin winter wheat lines.

KEYWORDS

- **Chemical mutagenesis**
- **Chromosome aberrations**
- **Electrophoresis**
- **Gliadins**
- **Molecular markers**
- **Mutations**
- **Polyacrylamide gel**

REFERENCES

1. Vavilov, N. I.; Theoretical Basevof Selection. Moscow: "Nauka" (Science); **1987**, 512 p. (in Russian).
2. Vasylkivskiy, S. P.; Forms creation and selection in posterity of winter wheat geneticalli non-stable mutants. Vasylkivskii, S. P.; Genetics and Selection in Ukraine at the Turn of the Millenium. In: Ed. Morgun, V. V.; (Chief ed.) et al., Kyiv: "Logos"; **2001**, *4(2)*, 207–211 (in Ukrainian).
3. Morgun, V. V.; Spontaneous and induced mutation changeability and its use in plants selection. Morgun, V. V.; Genetics and Selection in Ukraine at the Turn of the Millenium: In: Ed. Morgun, V. V.; et al. Kyiv: "Logos"; **2001**, *4(2)*, 144–174 (in Ukrainian).
4. Morgun, V. V.; Mutation Selection of Wheat. Morgun, V. V.; Logvinenko, V. F.; Kyev: "Naukova dumka" (Scientific Thought); **1995**, 627 p. (in Russian).
5. Vavilov, N. I.; Scientific Grounds of Wheat Selection. Moscow-Leningrad: "Selkhozhyz"; **1935**, 243 p. (in Russian).
6. Dubinin, N. P.; Genetical Principles Plants Selection. Moscow: "Nauka" (Science); **1971**, 7–32 (in Russian).
7. Zhuchenko, A. A.; Ecological Genetics of Crops. Kishinev: "Shtiintsa"; **1980**, 588 p. (in Russian).
8. Shcherbakov, V. K.; Mutations in Plants Evolution and Selection. Moscow: "Kolos"; **1982**, 327 p. (in Russian).
9. Dubinin, N. P.; Genetics. Kishinev: "Shtiintsa"; **1985**, 398 p. (in Russian).
10. Nadson, G. A.; On radioation impact on yeast fungus regarding general problem of radium impact on a live substance. *Bull. Roentgenol Radiol.* **1922**, *1*, 45–137 (in Russian).
11. Delone, L. N.; Mutation changeability role in plants selection. Bul. "VASHNIL" (All-union agricultural academy of the name Lenin). **1936**, *12*, 12–19 (in Russian).
12. Rapoport, I. A.; Carbonile Compounds and Chemical Mutations. Academy of Siences USSR Press; **1946**, *1(54)*, 65 – 68 (in Russian).
13. Rapoport, I. A.; Acetylation of gene molecules and mutations. Academy of Siences USSR Press; **1947**, *1(58)*, 119–122 (in Russian).
14. Rapoport, I. A.; Alkilation of Gene Molecules. Academy of Siences USSR Press; **1948**, *1(59)*, 1183–1186 (in Russian).

15. Rapoport, I. A.; Ethylen Oxide, Glicide and Glicole Impact on Gene Mutations. Academy of Siences USSR Press; **1948,** 60, 469–472 (in Russian).

16. Auerbah, Sh.; Problems of Mutagenesis: Translation from English. Moscow: "Mir"; **1978,** 463 p.

17. Williams, W.; Genetical Principles and Plant Breeding. Oxford: "Blackwell Scientific Publications"; **1964,** 448 p.

18. Kozachenko, M. R.; Experimental mutagenesis in barley selection. Kharkov: Institute of Crops Breeding of the Name Yuryev, V. Ya. NAAS; **2010,** 296 p. (in Ukrainian).

19. Kozachenko, M.; Mutagenesis in spring barley selection. Induced mutagenesis in splants selection: Coll. Sc. Papers. Bila Tserkva. - Institute of plants physiology and genetics of NAAS, Ukr. Soc. of geneticians and selectionists Vavilov, N. I. Bila Tserkva National Agrarian Universyty. **2012,** *188,* 195 (in Ukrainian).

20. Alekseeva, E. S.; Greenflower Buckwheat—its Present and Future. Alekseeva, E. S.; Kushny, V. P.; Kamyenets-Podilskiy: "Medobory (Moshak M. I.)" **2003,** 176 p. (in Ukrainian).

21. Alekseeva, O. S.; Genetics, selection and seedbreeding of buckwheat. Alekseeva, O. S.; Taranenko, L. K.; Malyna, M. M.; Kyiv: "Higher school"; **2004,** 213 p. (in Ukrainian).

22. Morgun, V. V.; Experimental mutagenesis and its use genetic improvement of crops (30 years reserch summary) Morgun, V. V. Physiology and Biochemistry of Crops. **1996,** *28(1–2),* 53–71 (in Russian).

23. Vasylkivskyi, S. P.; Mutation selection in the light of ideas of I. A. Rapoport. Indused Mutagenesis in Crops Selection: Collection of Scientific Papers. Bila Tserkva: Institute of Plants Physiology and Genetics of NAAS, Vavilov Ukrainian Association of Genetics and Breeders, Bila Tserkva National Agrarian Universyty; **2012,** 30–37 (in Ukrainian).

24. Vancetovič J.; Mladenovič-Drinich S.; Babič M.; Ignjatovič-Micich D.; and Andelkovič V. Maize genebank collections as potentially valuable breeding material. Genetika, 2010. Vol. 42, № 1, 9–21.

25. Vasylkivskyi, S. P.; and Vlasenko, V. A.; Genetic diversity enlargement of initial material in grains selection. Vasylkivskyi, S. P.; Vlasenko, V. A.; Sc.-Techn. Bul. of Myronivka Institute by V. M. Remeslo. Agrarian Science; **2002,** *2,* 12–17 (in Ukrainian).

26. Soyfer, V. N.; Moleculyar Mechanisms of Mutagenesis. Soyfer, V. N.; Moscow: "Nauka" (Science); **1969,** 511 p. (in Russian).

27. Rapoport, I. A.; Mechanism of mutation effects of N-nitrose compounds and the rule of mutagenes direct action. Chemical Mutagenesis Discovery: Selected Papers. Moscow: "Nauka" (Science); **1993,** 257–261 (in Russian).

28. Kozub, N. A.; Locuses of reserve proteins of soft wheat as markers of locuses of quantity features. Kozub, N. A.; Sozinov, I. A.; Faktors of Experimental Evolution of Organisms: Col. Sc. Papers. Kharkov: Agrarian Science; **2003,** 387–391 (in Russian).

29. Pomortsev, A. A.; Identification and estimation of barley seeds sort purity by method of electroforetic analysis of grain reserve proteins. Pomortsev, A. A.; Lyalina, E. V.; Moscow: Moscow Agricultural Academy; **2003,** 84 p. (in Russian).

30. Lisnevych, L. O.; Characteristics of geographical demarcation sorts of selection of Institute of plants physiology and genetics of NAAS by locuses of highmoleculyar glutenes and gliadins. Lisnevych, L. O.; Logvynenko, V. F.; Zlatska, A. V. Factors of Experymental Evolution of Organisms: Collective Science Papers. Ed. Roik, M. V.; Kyiv: Agrarian Science; **2003,** 392–397 (in Ukrainian).

31. Konarev, V. H.; Morphogenesis and Moleculyar-Biological Analysis of Crops. SPb: VIR; **1998,** 370 p. (in Russian).

32. Vasylkivskii, S. P.; Peculiarities of Applying Chemical Mutagenesis at Creating the Initial Material in Wheat Selection: Abstr. of DSc of Agricultural. 06.01.05 Odessa: Selection and seedbreeding. **1999,** 40 p. (in Ukrainian).

33. Khomenko, T. M.; Initial Material Creation in Winter Wheat Selection on the Basis of Induced Mutations: Abstr. for PhD. 06.01.05 Odessa: Plants Selection. **2006,** 20 p. (in Ukrainian).

34. Khomenko, T. M.; Identification of genotypes of mutant lines of winter wheat by gliadines. Induced mutagenesis in plants selection: Collective Science Papers. Bila Tserkva. Institute of Plants Physiology and Genetics of NAAS, Ukr. Soc. of Geneticians and Selectionists by Vavilov, N.I. Bila Tserkva National Agrarian Universyty; **2012,** 74–83 (in Ukrainian).

CHAPTER 24

ABOUT CYTOGENETIC MECHANISM OF CHEMICAL MUTAGENESIS

LARISSA I. WEISFELD

CONTENTS

24.1 INTRODUCTION

24.1.1 A SUMMARY ABOUT HISTORY OF CYTOGENETIC RESEARCH STUDIES OF CHEMICAL MUTAGENESIS

Chemical mutagenesis was discovered in the 1940s of the twentieth century [1–3]. The scientists of many countries began to actively study the cytogenetic effects of mutagens in meristem of growing tissues of plants. There was the openly delayed and not delayed effect of mutagens and fundamental distinction of radiation and chemical mutagenesis on cytological level (see e.g. reviews 4,5). Aberrations in ana- and telo-phases in plants were studied [3–5].

Many research studies were carried out on plants of *Crepis capillaris* L. This species has three pairs of the distinguishable chromosomes. Else until the discovery of the chemical mutagenesis, Navashin [6] investigated the karyology of genus of *Crepis.*

Chemical mutagens have the delayed effect on cellular level, that is, they cause damages of chromosomes only during the synthesis of DNA. After the termination of DNA synthesis—in the postsynthetic phase of G_2 and before its beginning—in phase of G_1, the chromosomes are not damaged regardless of the presence of chemical mutagens [4, 5]. A basic question consists in determination of mechanism of chemical mutagenesis on cellular level.

In the USSR, the study of mechanisms of chemical mutagenesis on cytogenetic level were performed in the laboratory of N.P. Dubinin with coworkers and by teams led by B.N. Sidorov and N.N. Sokolov.

Dubinin et al. (see review [7]) studied the cytogenetic effects of ionizing radiation, chemical compounds, in particular, of the alkylating agents. This group published a large number of papers and several monographs. N.P. Dubinin formulated the idea of the mechanism of the chemical mutagen effect: mutagens induce potential changes in DNA of chromosomes at all stages of the mitotic cycle, which come to light during a number of cell generations. He called it "chain process" in mutagenesis [8].

Sidorov and Sokolov [9, 10] had another point of view. In detail, their experiments are described in the following section. These authors analyzed the rearrangements of chromosomes in the plantlets of *C. capillaris*, that is, on the asynchronous populations of cells with regard to phases of mitotic cycle—G_1, S, and G_2.

24.2 AIM OF THE STUDY

In our work described later, we used a synchronous object, namely dry seeds of *C. capillaris*, whose bioblasts were supposed to be in the stage G_1. Effect of the chemical mutagen phosphemid was analyzed in comparison with the effect of X-rays. Alkylating agent phosphemidum (synonym phosphemid, phosphazin) is di-(etilene imid)-pyrimidyl-2-amidophosphoric acid. This compound is interesting by the fact, that it contains two ethylene imine groups and pyrimidine [11, 12]. We were studying the rearrangements of chromosomes in metaphases of plantlets of *C. capillaris* (L.).

24.3 MATERIALS AND METHODOLOGY

In experiments with phosphemid, air-dried seeds of *C. capillaris* of yield of 1967 were analyzed in 1968: in April (8 months of storage after harvest), June (10 months of storage), and July (11 months of storage). Distilled water was used both in the control and experiments. Chromosome rearrangements were analyzed in plantlets (meristem of tip roots).

In control and experiments, we used distilled water.

A total of 100 seeds were treated in an aqueous solution of phosphemid in the following concentrations: 1×10^{-2} M (22.4 mg was dissolved in 10 ml water), 2×10^{-2} M (22.4 mg was dissolved in 50 ml), or 2×10^{-3} M (2.24 mg dissolved in 50 ml of water). The treatment was carried out at room temperature (19–21°C) during 3 h. Then, seeds, which were treated by the mutagen, were washed in running water. The washed seeds were placed in Petri dishes on filter paper moistened with the solution of colchicine (0.01%). Seeds germinated in a thermostat at 25 °C, but in July 1968 because of the hot weather, the temperature in the thermostat could reach 27 °C. In parallel control experiments, the seeds were treated with aqua distillate.

After 24, 27, 31, and 36 h after soaking of seeds, we selected plantlets, took away plantlets not longer than 1 mm, and named them plantlets. The term *plantlet* refers to those plantlets that emerge after the beginning of soaking of dry seeds. These plantlets were placed in a Petri dish for further germination and subsequent fixation.

Root tips were cut off with a razor and placed in the following solution: 96 percent ethanol 3 parts + 1 part of glacial acetic acid. Solution was poured out after 3—4 h. Plantlets were washed for 45 min in 70 percent alcohol for keeping. We prepared temporary pressure preparations: fixed root tips were stained by acetous carmine and crushed in a solution of chloral hydrate between the slide and cover slip. We analyzed chromosome aberrations in metaphase plates of plantlets in the first division after treatment of seeds ($2n$-karyotype). In each plantlet, all metaphases were counted. Intact seeds of yields of 1966, 1067, and 1969 were used as control. These plantlets were fixed at different time intervals from 3 to 24 h.

Mitotic activity in the plantlets was detected on metaphase plates, depending on the number of nuclei in plantlets in the control and experiment. We used the plantlets in different intervals of time after the plantlet. In each plantlet, we counted between 500 and 1000 nuclei. We calculated the ratio of amount of plantlets with metaphase to all observed plantlets at all stages of fixation.

In all experiments, standard deviation from the mean was estimated.

In experiments with irradiation, seeds were exposed to X-rays in doses of 3,000; 2,000; and 4,000 R. Plantlets were selected for further germination for 22, 24, and 27 h after their soakage, fixed them, and the preparations were made, as described above, 0, 2, 3, 6, and 9 h after appear of plantlets. Control tests were performed in parallel and their results were combined.

24.4   RESULTS AND DISCUSSION

Radiation treatment by X-rays in the doses of 2000 and 3000 R evoked high percent of rearrangements of chromosomes in the early plantlets (Table 24.1).

Ecological Consequences of Increasing Crop Productivity

TABLE 24.1 Rearrangements of chromosomes in the early plantlets of *Crepis capillaris* after radiation treatment of seeds by X-rays

Dose of irradiation (K)	Time (h) From soakage to plantlets	From an arising of plantlets to fixation	General number of plantlets with metaphases	Number of metaphases Σ	With rearrange-ments Σ	(%)	Number of rearrangements Σ	For 100 metaphases	+m (%)
3000	22	0*	5	50	39	78.0	50	100.0	22.0
	24	3	16	334	179	53.6	203	60.8	7.2
		6	13	177	64	36.2	68	38.4	2.2
	27	2	21	184	96	52.2	123	66.8	14.6
		6	11	273	104	38.1	124	45.4	7.3
		9	9	377	101	26.8	106	28.1	1.3
	38	0*	8	96	35	36.4	38	39.6	3.2
2000	24	4	19	134	53	39.7	61	45.5	5.8
	29	3	22	229	58	25.3	61	26.7	1.4
4000	Growth is delayed by 30–39 h								
Control			18	1316	39	2.96	39	2.96	0.0

Note. *—plantlets less than 1 mm length were fixed.

The highest percentage of rearrangements, 60.8 and 45.5 percent were observed in the plantlets selected 24 and 27 h after a soakage, respectively.

About 24 and 27 h after X-ray irradiation of seeds, the frequency of rearrangements of chromosomes in plantlets started from maximum. This frequency is decreasing with increasing time from arising of plantlet (see Table 24.1). About 22 h after irradiation in plantlets fixed in 0 h, only five plantlets with metaphases were observed. Their level of rearrangements of chromosomes was high: 50 metaphases contained 78 percent of the rearrangements; 38 h after the treatment eight plantlets were plantlet and containing 364 percent of metaphases with rearrangements. In all the variants of plantlets and fixation, the frequencies of rearrangements per 100 cells were higher than those of metaphases with rearrangements, which means that it was more than one rearrangement per cell (see column of m, % in Table 24.1).

Under exposure to radiation in the dose of 4000 R, the germination of seeds was significantly delayed. In metaphases after radiation of seeds, rearrangements of chromosomal type prevailed. Rearrangements of chromatid type were also observed, but their frequency did not differ significantly from control (Table 24.2).

TABLE 24.2   Chromosomal and chromatid types of rearrangements in control experiment

Number of plantlets	Number of metaphases	Number of rearrangements (%)	Chromatid rearrangements (%)	Rearrangements of chromosomal type (%)
117	1316	2.96	2.58	0.38

Similar relationship between the number of rearrangements and the number of metaphases with rearrangements under the radiation treatment of plantlets was observed by Rapoport [13]. In this study, frequency of rearrangements also decreased in the seeds with later terms of germination after a treatment with X-rays.

Phosphemid, on the contrary, caused only chromatid rearrangements in all variants of the experiment. Under moderate concentrations of phosphemid, the frequency of rearrangements in one cell did not exceed the frequency of metaphases with rearrangements. Figure shows that phosphemid in a moderate dose (2 ... 10^{-3} M) causes rearrangements of chromosomes in the plantlets of first plantlet—24 h after the soakage of seed. Their frequency increases with the increase of term of fixation.

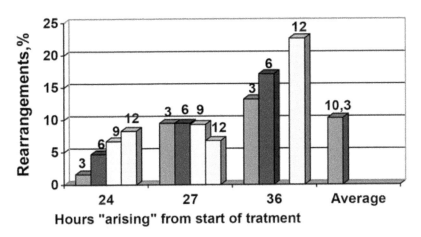

FIGURE 24.1   Rearrangements of chromosomes in early plantlets after the treatment of seeds by phosphemid in a concentration 2×10^{-3} M

On a y-axis—rearrangements, %. X-axis represents the time after the soakage of seeds: 24, 27, and 36 h. Plantlets were fixed 3, 6, 9, or 12 h (numbers above columns) after plantlet originates. Data are found to be statistically significant.

The later plantlet—27 h after soaking was characterized by the more high frequency of rearrangements than upon the fixation—12 h in first plantlet—24 h. After 36 h after soaking of seeds, the frequency of rearrangements increased again in comparison with last fixation (27 h after soaking of dry seeds) (see a Figure 24.1).

All rearrangements were of the chromatid type. They arise during the synthesis of DNA.

On the basis of these facts, we conclude that mutagen was presented in cells during all periods of fixation, affecting chromosomes after their passing to the DNA synthesis and remained in cells 24, 27, and 36 h after plantlet, in spite of washing of seeds by tap water from the phosphemid. Increase number of rearrangements under fixation 2, 6, 9, and 12 h after arising of plantlet in 24 h after treatment indicates the heterogeneity of phase of G_1 in dry seeds. It is not a "delayed effect" of chemical mutagens, typical for asynchronous population of cells, when chemical mutagens do not cause rearrangements of chromosomes at G_2 or G_1 of mitotic cycle. In our experiments, the growth of number of the rearrangements in plantlets can result from the storage of mutagen in cells and gradual transmission of cell cycle from the G_1 stage to the synthesis of DNA. Increase of number of rearrangements in the plantlet after 36 h can be explained also by the lack of synchronization of passage of cell cycle to the synthesis of DNA by different cells in the presence of mutagen. Table 24.3 and a figure show the mitotic activity (number of the mitoses in plantlets) at different periods after plantlet or of fixation.

TABLE 24.3 Mitotic activity of cells in plantlets of *Crepis capillaris* after treatment of air-dried seeds with phosphemid at a concentration 2×10^{-3} M. Harvest 1967. Analysis 1968

Time from the beginning of treatment (h)	Time from plantlet to fixing (h)	Plantlets, number		
		Total	With mitoses	(%)
April				
24	3	57	24	42.1
-	6	56	32	57.1
-	8	54	36	66.7
27	3	47	16	35.5
-	5	19	12	65.0
On average		233	120	51.5 ± 3.28
June				
24	3	38	5	13,2
-	6	28	16	57.1
Time from the beginning of treatment (h)	Time from plantlet to fixing (h)	Plantlets, number		
		Total	With mitoses	(%)

TABLE 24.3 *(Continued)*

-	9	19	15	78.9
-	12	27	23	85.2
27	3	35	8	22.9
-	6	35	17	48.6
-	9	23	12	52.2
-	12	23	20	87.0
36	3	37	22	59.5
-	6	22	10	45.5
-	12	20	15	75.0
On average		307	163	53.1 ± 2.85
July				
24	3	43	21	48.8
-	4	30	15	50.0
-	5	26	22	84.2
-	7	17	13	76.5
-	9	18	13	72.2
27	3,5	34	13	78.2
-	5	44	31	70.5
-	6	36	22	61.1
-	7	26	24	92.3
-	8	22	20	91.0
31	3	34	23	67.6
-	4	59	33	55.9
average		289	360	64.3 ± 2.43

On an average, the mitotic activity was increasing in senescent seeds during 1968 from 51 percent in April to 64 percent in July 1968 (see Table 24.3). Mitotic activity within the limits of every plantlet also increased during a year: April, June, and July (see Figure 24.2, Table 24.3). Frequency of rearrangements also increased (see Figure 24.2).

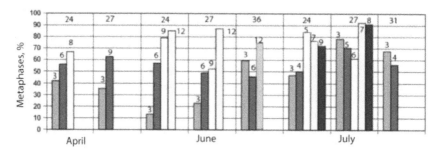

FIGURE 24.2 Mitotic activity in the cells of plantlets of *Crepis capillaris* after treatment of seeds by phosphemid in a concentration 2×10^{-3} M. Harvest 1967. Plantlets took away for sprouting at 24, 27, 31, and 36 h after their appearance. Then, plantlets analyzed at different times after that during 3–12 h (numerical symbols over columns). Analysis was performed in April, June, and July 1968.

The time after the soakage of seeds was 24, 27, and 36 h. Plantlets are fixed at 3, 4, 5, 6, 7, 8, 9, and 12 h (numerals over columns) after arising of plantlet. Data are statistically significant.

Thus, the later the cell enters into mitosis, the longer the synthesis of DNA will undergo the effect of mutagen in cells.

The high dose 1×10^{-2} M of phosphemid is injuring the chromosomes, considerably stronger on the background of the decline of number of plantlets with metaphases. In 72 explored plantlets, 771 metaphases were observed; in 166 (22%) of them, rearrangements were observed. In 22 metaphases, plural breakages of chromosomes were observed. Frequency of plural breakages increased with the increase of term of fixation. 24 h after plantlet more than 21 percent metaphases contained plural breakages of chromosomes.

Besides the large number of breakages of chromosomes the high doses of phosphemid resulted in abnormality of spindle of division. It also confirms that mutagen is interlinked with the proteins of cytoplasm.

Rapoport, discoverer of the method of chemical mutagenesis, proved that chemical compounds—ketonic connections, aldehydes, ethylene imine, and others interact with the proteins of cytoplasm and chromosomes. His experimental data and theoretical conclusions were published in a number of articles, which now are united in the edition of the "Discovery of Chemical Mutagenesis" [13].

Probably, phosphemid cooperates with proteins by means of two groups of ethylene imine. In addition, the pyrimidine in its structure can be included in the structure of chromosomes at the time of the synthesis of DNA. Moreover, phosphemid not only causes rearrangements of chromosomes, but it is also repress mitosis in cells, and id est. affects proteins of spindle of cell division. It becomes apparent under the influence of the high doses of phosphemid, leading to plural chromosomal breakages and the conglomeration of chromosomes in the form of "stars" in the center of cell. The authors of Ref. [14] determined the number of nuclei labeled by H^3-thymidin and the number of labeled mitosis in plantlets after treatment of dry seeds by H^3-thymidin through 24, 30, and 36 h after its introduction. The first labeled nuclei appeared 10 h after beginning of labeling. These data confirm heterogeneity of phase of G_1 in dry seeds.

In the research [8], Sidorov, Sokolov, and Andreev have processed the plantlets of C. capillaris by ethylene imine. Plantlets were grown in the presence of colchicine. During the five generations of polyploid cells, the doubled rearrangements of chromosomes were observed. Also in polyploid, cells were plantlet new rearrangements of chromatid type, undoubted. In the study [9], plantlets were treated by ethylene imine, then washed in tap water, and then cut into very small fragments in the form of "gruel." Intact plantlets were treated by this "gruel," washed in the water off and placed in Petri dish for sprouting in solution of colchicine. In metaphases of plantlets treated by "gruel," the authors detected new aberrations of chromosome of chromatid type. This indicates the preservation of mutagen in cells treated by this "gruel." These authors named the appearance of new rearrangements in a series of generations of cells "secondary mutagenesis."

In the article given in Ref. [9] authors studied the effect of ethylene imine on dry seeds. Rearrangements only of chromatid type were observed. It shows that chromosomes in the G_1 are not sensitive to the chemical mutagen. Ionizing radiation, vice versa, damages chromosomes at all stages of cellular cycle.

However, here growth of frequency of alterations after the treatment of seeds by ethylene imine is not observed, in contrast to our data (see Table 24.1, Figure 24.2). Frequency of rearrangements had as tough wavy character (Table 24.4).

TABLE 24.4 Effect of ethylene imine (solution 0.1%) on the dry seeds of Crepis capillaris (from Ref. [9])

Time of fixing (h)	Number of cells	Rearrangements	
		Number	(%)
2	554	36	6.50 ± 1.05
8	672	23	3.42 ± 0.70
12	1,018	88	8.64 ± 0.88
16	1,000	38	$3.80 \pm 0,61$

There was no increase of number of alterations in dependence on time of fixing. Probably, here plantlets were fixed later, then in my experiments, length of plantlets varies. According to our observations, even 2 h after plantlet, the length of plantlets varied. Wavelike character of effect of ethylene imine in Ref. [9] also shows heterogeneity of phase of G_1 in dry seeds. Maybe the synthesis of DNA under the mutagen impact is yet more asynchronous.

The phenomenon of preservation of chemical mutagen in the cells of plants has a practical value. Damages of chromosomes remained in cells are transmitted after division and organogenesis of any plant. It is necessary to take into account genetic harm of herbicides of type 2,4-D, of izo-proturon, [15] and of pesticides [16] in the same year after their use. It is necessary also to take into account chemical contamination of environment [17].

24.5 CONCLUSIONS

Our data and results reported in the literature show that chemical mutagens penetrate in the cells and remain in them, probably in the form of compounds with proteins. Thus, the synthesis of DNA and chromosomes in cells is damaged. The rough damages of chromosomal structure are not preserved in the new cells after division. Such cells perish. The mutagens from environment can cause insignificant changes in the structure of DNA and chromosomes. Such damages are transmitted to the subsequent generations of cells after their division in a growing organ. In the analysis of effect of chemical mutagens from the environment on cells, it is needed to take into account inevitability of their presence in a cytoplasm and kernel of cells.

Although radiation causes the direct breaking of chromosomes, after an ionizing radiation, secondary mutagens can also appear, if the radiation breaks the structure of cell proteins.

ACKNOWLEDGEMENTS

Apart from other geneticists and plant breeders of the former USSR, I am thankful to Iosif Abramovich Rapoport—he was a prominent scientist who lived a heroic life. Even now in our everyday work, we use the chemical mutagens, which he discovered and synthesized more than 50 years ago as well as his methodical advice.

Special thanks goes to Alexander V. Bukhonin for help in preparations of illustrations.

KEYWORDS

- **Chemical mutagenesis**
- **Colchicine**
- **Ethylene imine**
- **Plantlet**
- **Rearrangement**
- **Seed**
- **Seeds**
- **X-rays**

REFERENCES

1. Rapoport, J. A.; Ketonic Connections and Chemical Mechanisms of Mutations. Doklaly Academy of Siences USSR (Press); **1946**, *54(1),* 65–68. (in Russian).
2. Auerbach, Ch.; and Robson, J. M.; Production of mutations by allyl isothiocyanate. *Nat.* **1944,** *154,* 81.

3. Auerbach, Ch.; and Robson, J. M.; Chemical production of mutations. *Lett Ed. Nat.* **1946,** *157,* 302.
4. Kihlman, B. A.; Aberrations induced by radiomimetic compounds and their relations to radiation induced aberrations. Radiation-Induced Chromosome Aberrations. Ed. Wolff, S.; New York–London: Columbia University Press; **1963,** 100–122.
5. Loveless, A.; Genetic Allied Effects of Alkylating Agents. London. 1966. Translation Ed. Dubinin's, N. P. Moscow: "Nauka"; **1970,** 255 p. (in Russian).
6. Navashin, M. S.; Problems Karyotype and Cytogenetic Studies in the Genus *Crepis*. Moscow: "Nauka"; **1985,** 349 p. (in Russian).
7. Dubinin, L. G.; Structural Mutations in the Experiments with *Crepis Capillaris.* Moscow: "Nauka" (Science); **1978,** 187 p. (in Russian).
8. Dubinin, N. P.; Some Key Questions of the Modern Theory of Mutations. *Genetika (Genetics).* **1966,** *7,* 3–20. (in Russian).
9. Sidorov, B. N.; Sokolov, N. N.; and Andreev, V. A.; Mutagenic effect of ethylene imine in a number of cell generations. *Genetika (Genetics).* **1965,** *1,* 121–122. (in Russian).
10. Andreev, V. S.; Sidorov, B. N.; and Sokolov, N. N.; The Reasons for Long-Term Mutagenic Action of Ethylenimine. *Genetika (Genetics).* **1966,** *4,* 28–36. (in Russian).
11. Chernov, V. A.; Cytotoxic Substances in Chemotherapy of Malignant Tumors. Moscow: Medicine; **1964,** 320 p. (in Russian).
12. Weisfeld, L. I.; Damages of chromosomes and divisions of mitotic activity in plantlets of *Crepis capillaris* by alkylating antineoplastic agent. *J. Inf. Int Knowl.* **2012,** *4(4),* 295–307.
13. Rapoport, J. A.; Discovery of Chemical Mutagenesis. The Chosen Works. Moscow: "Nauka" (Science); **1993,** 304 p. (in Russian).
14. Protopopova, E. M.; Shevchenko, V. V.; and Generalova, M. V.; Beginning of DNA synthesis in seeds *Crepis capillaris. Genetika (Genet.).* **1967,** *6,* 19–23. (in Russian).
15. Sanjay Kumar. Effect of 2,4-D and Isoproturon on chromosomal disturbances during mitotic division in root tip cells of *Triticum aestivum* L. *Cytol Genet.* **2010,** *44(2),* 14–21.
16. Pandey, R. M.; Cytotoxic effects of pesticide in somatic cells of *Vicia faba* L. *Cytol. Genet.* **2008,** *42(6),* 13–18.
17. Lekjavichjus, R. K.; Chemical Mutagenesis and Contamination of Environment. Vilnjus: Moksklas; **1983,** 223 p. (in Russian).

CHAPTER 25

IMPORTANCE OF DISCOVERY OF I. A. RAPOPORT OF CHEMICAL MUTAGENESIS IN THE STUDY OF MECHANISM OF CYTOGENETIC EFFECT OF MUTAGENS

LARISSA I. WEISFELD

CONTENTS

25.1 INTRODUCTION

Diversity of cultural plants is possible to be obtained by the induction of genetics or in other words by hereditary changes. For achieving a genetic variety, use of chemical mutagens is necessary. The search of chemicals that are able to cause the hereditary changes began both abroad and in us after the invention of mutagenic action of the X-rays by Müller [1] N.K. Kolzov is a prominent Russian biologist—who is a geneticist who predicted the idea of matrix reproduction of chromosomes. He was the first to propose that mutations could arise up not only from an ionizing radiation but also under the action of chemicals.

Then, Saharov [2] by adding to larvae and pupas of *Drosophila* a solution of iodine he was able to sex-link mutations in the second generation of flies. Analogical experiments were carried out with ammonia by Lobashov [3,4]. Ammonia did not cause mutations. The level of mutations of adult individuals under the impact of the iodine or acetic acid exceeded a natural level just a bit. In 1939, Gershenson published a work about induction of mutation in *Drosophila melanogaster* by the action of sodium salt of timonucleic acid [5].

In 1946, in England Charlotte Auerbach and Joule Robson published short reports about induction of mutations among adult males of *Drosophila* after affecting larvae containing poison substance of mustard gas (allyl isothiocyanate—mustard gas) [6,7]. Authors discovered that 24 percent are sex-link mutations. A great geneticist Joseph (Iosif—By Russian language literature) Abramovich Rapoport worked out the theory and method of chemical mutagenesis as a phenomenon. In the opinion of Rapoport [8], in the earlier mentioned works, there was low mutational activity and convincing data were not represented about the effects of the tested substances and their genetic effects.

25.2 A WORLD FAMOUS SCIENTIST J. A. RAPOPORT AND THE PHENOMENON OF FORMATION IN CELLS OF THE SECONDARY MUTAGENS

Rapoport began the experimental search of substances which are able to induce mutations; initially he was a participant of Great patriotic war. After his return from army, he published an article in 1946 [8]. He showed the origin of mutations in adult individuals of *Drosophila* after treatment of larvae and eggs by formaldehyde, aldehydes, urotropine, and other ketonic connections. In that investigation [8], high frequency of lethal mutations analyzable on the method of *ClB* (estimation of lethal mutations for adult individuals with the marker sign of eyes as stria of Bar) was obtained. Formaldehyde when you add at nutrient larvae medium induces 6 percent, and in some experiments 12 and 30 percent of lethal and visible mutations against less than 0,12 percent in the control. As well as after a short-wave radiation, ketonic connections caused the "tufts of similar changes " Subsequent works of Rapoport were completed by creation of the theory of chemical mutagenesis. Rapoport marked the fundamental difference of action of short-wave radiation as of mutagen from the action of chemical mutagens [9]. A radiation causes the break in chromosomes with subsequent confluence of the torn fragments in different combinations; as a result dominant mutations, many arise. Mustard gas operates similarly [9]. Rapoport [6,9–11] explains the action of ketonic connections and alkylating agents, when they penetrate in the cells, where they interact with amino groups, what results in the irreversible changes of proteins of cytoplasm and genic proteins. Thus recessive mutations appear mainly unlike the breaks of chromosomes after a short-wave radiation [12]. In experiments with addition of chemical agents to the nutrient medium of culture of drosophila, the mutagen is present there constantly. Rapoport supposed that the complex formation of chemicals with proteins in cytoplasm and chromosomes determines the mutagenicity of these connections.

This capacity of complex formation of chemical mutagens with proteins in a cell predetermined the preservation of mutagens and caused appearance of mutations or transpositions in a row of cellular generations.

Considerably later classic works of B.N. Sidorov and N.N. Sokolov on the seedlings of well-known cytogenetic object of *Crepis capillaris* have showed that a mutagen an ethylene imine causes new rearrangements of chromosomes in row of cellular generations [13,14]. After treatment of seedlings, the mutagen were washed off by tap water. These authors also showed that ethylene imine interacts in the cells with biologically important substances, such as amino acids or the predecessors of DNA [15]. These complexes are named as secondary mutagens by the authors. Hence, the appearance of the new rearrangements of chromosomes in cell generations is named as "the secondary mutagenesis."

In the research studies [16,17], I.A. Rapport in detail analyzed the objective laws of origin of the lethal mutations from alkylating agents, likeliness of effect of chemical and radiation mutagens, and fundamental distinction of the chemical mutagenesis from the effect of the ionizing radiation. He made it clear that the supermutagens [18] were provoking a large number of heritable changes. Research studies of Rapoport have resulted in the method of chemical mutagenesis, which finds its application in genetics, cytogenetics, and agriculture. Since 1959 Rapoport published collections of scientific research studies of selectionists from the entire country. It is possible to name last [19,20]. This method is used in large scientific collectives both in our country and abroad. It also has application in experimental biology, microbiology, agriculture, and medicine. By means of chemical mutagens the registered sorts of agricultural cultures were created. For example, the cultivars of oat and barley were created [21]. Eiges with collaborators created cultivars of winter wheat, which are the Sibirskaja niva (Siberian field), the Stavropolskaja kormovaja (Stavropol fodder), and the Imeni Rapoporta (Name of Rapoport) [22] and perspective specimens, so plant-breeding work proceeds with that.

Collections of research studies of Rapoport [16,17], books on his activity [23,24], and his fundamental theoretical works [25] were published posthumously. These publications were fulfilled with the initiative and the participation of his wife who is a doctor of biological sciences in O.G. Stroeva. To the century of A. Rapoport, conferences and publications were devoted. Research studies with the use of chemical mutagenesis methods continued, in particular in works of Eiges et al. [26], which have large achievements in genetics, selection, and practical applications of this method on the winter wheat mutants.

KEYWORDS

- Alkylating agents
- Chemical mutagenesis
- Ethylene imine
- Ketonic connections
- Mustard gas

REFERENCES

1. Müller, H. J.; The problem of genic modification. Ztschr. Ind. Abst. Vererb. (Zeitschrift Für Induktive Abstammungs Und Vererbungslehre). **1928**, B. *1,* Anb. 1, S. 234–266. (By German).
2. Saharov, V. V.; Iodine as chemical factor of influencing on a mutational process at *Drosophila melanogaster. Biol. J.* **1932,** *I(VIII)(3–4),* 1–8. (by Russian).
3. Lobashov, V. E.; and Smirnow, F. A.; To nature of action of chemical agents on a mutational process. Report of I. Action ac.a. on non-disjunction and transgenations. Doklady, A. N. (Report of Academy of Sciences) of the USSR; **1934,** *2(III)(5),* 307–311. (by Russian).
4. Lobashov, V. E.; and Smirnow, F. A.; To nature of action of chemical agents on a mutational process. Report of II. Action ammonia on the origin of lethal transgenations. Doklady, A. N. (Report of Academy of Sciences) of the USSR; **1934,** *3(IV)(3),* 174–178. (by Russian).
5. Gershenson, S. M.; Induction of the directed mutations at *Drosophila melanogaster.* Doklady, A. N. (Report of Academy of Sciences) of the USSR; **1929,** *25,* 224–227. (by Russian).
6. Auerbach, Ch.; and Robson, J. M.; Production of mutations by allyl isothiocyanate. *Nat.* **1944,** *154,* 81. (on English).
7. Auerbach, Ch.; and Robson, J. M.; Chemical production of mutations. *Lett. Ed. Nat.* **1946,** *157,* 302. (on English).
8. Rapoport, I. A.; Ketonic Connections and Chemical Mechanisms of Mutations. Doklady, A. N. (Report of Academy of Sciences) of the USSR; **1946,** *54(1),* 65–68. (by Russian).
9. Rapoport, I. A.; A Reaction of Free Amino-Groups in Genic Proteins. Doklady, A. N. (Report of Academy of Sciences) of the USSR; **1947,** *56(5),* 537–540. (by Russian).
10. Rapoport, I. A.; Chemical reaction with amino-group of protein in the structure of genes. *J. General Biol.* **1947,** *8(5),* 359–379. (by Russian).
11. Rapoport, I. A.; The alkylating of genic molecule. Doklady, A. N. (Report of Academy of Sciences) of the USSR; **1948,** *59(6),* 1183–1186. (by Russian).
12. Rapoport, I. A.; Interaction of Ethylene Imine with Genic Proteins and its Influencing on Heritable Changes. Bulletin of Moscow Society of Investigators of Nature. Biology. **1962,** *LXVII(1),* 96–114. (by Russian).
13. Sidorov, B. N.; Sokolov, N. N.; and Andreev, V. A.; Mutagenic effect of ethylene imine in a number of cell generations. *Genetika (Genet.).* **1965,** *1,* 121–122. (by Russian).
14. Andreev, V. S.; Sidorov, B. N.; and Sokolov, N. N.; The reasons for long-term mutagenic action of ethylenimine. *Genetika (Genet.).* **1966,** *4,* 28–36. (by Russian).
15. Sidorov, B. N.; Sokolov, N. N.; and Andreev, V. A.; Highly active secondary alkylating mutagens. *Genetika (Genet.).* **1966,** *7,* 124–133. (by Russian).
16. Rapoport, I. A.; Opening of chemical mutagenesis. The Chosen Works. Moscow: Nauka (Science); **1993,** 304 p. (by Russian).
17. Rapoport, I. A.; Genes, Evolution, Selection: The Selected Labors. Moscow: Nauka (Science); **1996,** 246 p. (by Russian).
18. Rapoport, I. A.; Features and mechanism of action of super mutagens. Supermutagens Ed. Rapoport, J. A. Moscow: Nauka (Science); **1966,** 9–23. (by Russian).

19. Chemical Mutagens and Pata-Aminobenzoic Acid in the Increase of the Productivity of Agricultural Plants. Moscow: Nauka (Science); Red, I. A. Rapoport. **1989,** 253 p. (by Russian).

20. Chemical Mutagenesis and Problems of Agricultural Production. Moscow: Nauka (Science); Red, I. A. Rapoport. **1993,** 238 p. (by Russian).

21. Shevtsov, V. M.; and Maljuga, N. G.; Selection and agro technics of barley on Kuban. Krasnodar: The Kuban State Agrarian University; **2008,** 138 p. (by Russian).

22. State Register of the Plant-Breeding Achievements Admitted to the Use. Moscow; **2013,** 5–8. (by Russian).

23. Josef Abramowitz Rapoport—scientist, warrior, citizen. Essays, memoirs, materials. Compiled by O.G. Stroeva. Moscow: Nauka (Science); **2003,** 335 p. (by Russian).

24. Stroeva, O. G.; Josef Abramowitz Rapoport, 1912–1990. Moscow: Nauka (Science); 215 p. (by Russian).

25. Rapoport; I. A.; Microgenetics. Moscow: Reprint Edition; **2010,** 530 p. (by Russian).

26. Eiges, N. S.; Historical role of Joseph Abramovich Rapoport in genetics. Continuation of researches with the use of method of chemical mutagenesis. *Vavilov J. Genet Selection.* **2013,** *17(1),* 162–172. (by Russian).

CHAPTER 26

THE USE OF THE REGULARITIES OF MANIFESTATION OF HEREDITARY VARIABILITY CHARACTERISTICS IN LEGUMES IN THE SELECTION OF *GALEGA ORIENTALIS*

VERA IV. BUSHUYEVA

CONTENTS

26.1   INTRODUCTION

Galega orientalis Lam. is a new promising leguminous crop with high genetic potential, which has important practical value in the intensification of agricultural production. Because of the lack of breeding study, its genetic potential is not yet fully revealed. The urgent problem in the selection of the crop is to obtain genotypic variability and breeding of new, patentable varieties [1].

Patentability of cultivars acquired urgency after adoption of the law in the Republic of Belarus "On Patents for Plant Varieties" (1995), which guarantees protection of copyright of the breeders at the state level [2]. After the entry of the Republic of Belarus in the International Union of states-parties on the Protection of New Varieties (UPOV) [3], such protection is made possible at the international level. By 2001, UPOV had been tasked to include in the number of protected all kinds of crops, including *G. orientalis* [4].

In connection with this, the problem of the development of criteria of novelty, distinctness, uniformity, and stability (DUS) became urgent, the solution of which is not possible without in-depth theoretical study on the selection of the crops and getting genetically diverse selection material. Belarus through membership in UPOV was able to not only use the joint experience of other states, but also to contribute to the development of the international plant breeding. The main problem in this work is to create a new gene pool and expand the spectrum of genotypic variability [5]. To solve this problem, we need theoretically substantiated and purposeful use of both natural and developed by breeding sources of genotypic variation [6].

Science has shown that genetic diversity of all cultivated plants was formed as a result of a long morphogenic process taking place under the influence of both natural factors and human activities. The most intensive morphogenic process in most species was in the center of their origin for a long historical period, which contributed to the formation of a wide range of variability traits, the nature of inheritance of which is reflected in the law of homologous series of N.I. Vavilov.

In the feed crops introduced recently, morphogenesis occurred in the wild populations less intensively and without human intervention, so their genetic potential is not fully disclosed yet [7]. The law of homologous series in hereditary variability, which, as noted M.E. Lobashov, "opens vast prospects of scientific knowledge and foresight" is of fundamental significance to speed up the process of selective improvement of feed crops and obtain a wide range of genotypic variability of traits (it is quoted on [8]).

The laws defined by N.I. Vavilov consist of a striking parallelism of the morphogenetic process in manifestation of genetic variation in closely related species, genera, and families. For example, for legumes, consistent and similar manifestation of variability in the flower color is theoretically substantiated, which is characterized by certain amplitude of variation and controlled by genes that have common evolutionary origin. As a result of study of the morphogenetic process in legumes (peas, lathyrus, gram, and beans), N.I. Vavilov found that primitive, mostly dominant forms, are distinguished by small steadfast seeds, beans are prone to splitting when ripe, are characterized by blue and often dark-blue flowers. In the selection improving process enlarged, white-seed, pink-flowering forms appear, and more highly cultivated forms have almost exclusively large white seeds and white-colored flowers, as N.I. Vavilov wrote, … "in the periphery there are often found extremely interesting original recessive forms as a result of inbreeding or mutational process" [7].

26.2 MATERIAL AND METHODOLOGY

As a white-colored flower is a sign of cultivated plants and is due to the mutation process that occurs in nature, the defined white-colored plant of *G. orientalis* in the many-year crops of the blue-flowering population of Nesterka variety was used as a source material for further breeding. Taking the law of homologous series in hereditary variation and defined in it parallelism of the inheritance of traits in legume crops as fundamentals, we have applied it in a practical selection of *G. orientalis* to obtain a wide range of its genotypic variability. For greater confidence in the correctness of the chosen area of research, the evolutionary path and the nature of the manifestation of the morphogenetic processes in different species of lupine (white, narrow-leaved, and perennial) were analyzed to identify the role of mutations in the variability of the flower coloring and the presence of parallelism in their inheritance. The analysis showed that in the most cultivated species of white lupine (*Lupinus albus* L.) after a long selection improvement, there are maximum recessive characteristics and properties of the mutant origin including white coloring of flowers, but within a species, there are forms with color ranging from dark-blue to white.

26.3 RESULTS AND DISCUSSION

The least selected species is perennial lupine (*Lupinus polyphilus* Lind.). Observations of the wild populations have shown that the color of its flowers is characterized by the same amplitude of variability as in other species, which confirms the parallelism in their evolutionary inheritance and common origin.

In the blue lupine (*Lupinus angustifolius* L.) at the initial stages of the introduction into culture, there were only blue flowering forms, which is why it got its name blue lupine. As a result of morphogenetic process that occurs in nature under the influence of spontaneous mutagenesis and selection improved by breeding and genetic techniques, the range of variability of all morphological features, including coloring of flowers, has grown considerably. There are forms of plants with dark-blue, blue, light-blue, lilac, pink, and white coloring of flowers [9].

The most selected white flowering variety of the blue lupine in our selection is Biser-347, it also combines a range of useful human recessive traits and properties of the mutant origin, such as white color of flowers and seeds, light-green without pigmentation vegetative organs, pods that do not split, does not contain alkaloid, and has determinant type of branching [10].

As a self-pollinated plant, blue lupine was a very convenient object of our research to study regularities of inheritance of traits and morphogenetic process in the hybrid generations. The laws defined by us [11] were used for scientifically substantiated prediction and acceleration of the selection process of *G. orientalis* and making in short time a new diversified initial breeding material [12, 13].

For this purpose, we have studied morphological traits of the wild populations of the selection variety samples of *G. orientalis* and identified natural parallelism manifestations of the hereditary changes in the coloring of flowers and other morphological and economically valuable traits. Thus, it was observed that in the centers of origin where *Galega* grows wild (Northern Caucasus, Dagestan, Georgia, Northern Armenia, and Southwestern part of Azerbaijan) typical for legume dominant features occur (blue color of flowers, green with pigmentation color of leaves, uneven ripening, etc.). In certain geographic areas (which is also confirmed by the law of homologous series) in plants of *G. orientalis*, new traits and properties appropriate to the specific growing conditions are formed. For example, to a wild population of *G. orientalis* from Northern Caucasus, the most characteristic traits are early maturity, early flowering and fruiting, high winter resistance, high seed production, rapid and uniform ripening, large leafiness, and plant height. For these traits, the best samples from this area were introduced into culture and on their basis, selection varieties that are being cultivated commercially now have been developed. As to the wild populations of *Galega orientals* from Northern Armenia, they are increasingly being used in breeding to enrich the gene pool and create initial material and varieties with certain distinctive features [13]. In the process of introduction, establishment, and expansion from sharply differing climatic conditions to others, *G. orientalis* showed a variety of new changes including white flowers. A white flowering spontaneous mutation found by us in the old-growth crops of *G. orientalis* was a confirmation of the presence of parallelism of variability of traits in legumes and was included in hybridization.

When forecasting a possible morphogenetic process in the hybrid generations of *G. orientalis*, examples with the blue lupine were used which revealed clear patterns of manifestation of the combination variability by this and other traits relevant to the law of homologous series.

To obtain natural hybrids, we left a white form in our common crops to full maturation of beans. It made it possible to obtain hybrid seeds from open pollination of its flowers with pollen of blue flowering plants surrounding it. The defined white flowering plant after harvesting the seeds was transplanted to an isolated plot with a view to its subsequent self-pollination and production of homozygous constant form that was subsequently subjected to selection elaboration and ended with the establishment of a constant variety sample with flowers and light-green vegetative organs.

The resulting hybrid seeds F_0 in the spring were used for planting nursery of the first-generation hybrids in order to further explore the morphogenetic process in the future hybrid progenies. In the first-generation hybrid plants of *G. orientalis*, according to the first law of Mendel, were characterized by uniformity and dark-green vegetative organs, indicating intermediate nature of inheritance of flower coloring and dominant manifestation of color characteristics of vegetative organs. Harvested in the nursery of the first-generation hybrids, the seeds were used for the nurseries of the second-generation hybrids. The nursery was laid by a wide cluster method to the wells with an area of nutrition 70 × 70 cm. This seeding method enabled to do a more accurate description of the obtained forms of the studied traits. The conducted phenological observations and a thorough analysis of each plant allowed defining the biotypes with a predicted combination of traits. In the hybrid population F_2, similar to the blue lupine, phenotypes with flower coloring varying from white to dark-blue appeared. Coloring of stems and leaves varied from light-green to dark-green and intense violet with pigmentation, leaf size–from small to large, the shape of the leaves—from lanceolate to elliptical and bush—from erect to semierect [12].

Combination variability emerged along the interphase periods, plant height and tillering, the elements of the structure of seed yield and mass of 1000 seeds (Table 26.1).

TABLE 26.1 Basic statistical characteristics of traits of *Galega orientalis* in hybrid population of F_2

Character	X_{min}	X_{max}	X_{av}	S	V (%)	S_x	S_x (%)
Phase of development from the beginning of growing (days)							
Budding	33	47	40	3.8	9.8	0.76	1.9
Florescence	42	52	47	2.2	4.7	0.44	1.0
Ripening	86	104	96	3.7	3.8	0.74	0.8
Other characteristics in terms of one plant							
Height (cm)	90	135	118	10.6	8.9	2.12	1.8
Stems (pcs)	7	23	14	3.0	21.5	0.60	4.3
Trusses (pcs)	38	104	70	15.4	22.0	3.1	4.4
Beans (pcs)	444	1756	1358	291.8	21.5	58.4	4.3
Seeds (pcs)	1490	7150	4322	1291	29.8	258.1	5.9
Seeds (g)	13.4	53.6	32.4	9.6	29.6	1.1	5.9
Seeds in a pod (pcs)	1.8	6.5	4.0	1.1	27.5	0.22	5.5
Mass of 1,000 seeds (g)	6.72	9.72	8.1	1.2	14.8	0.24	3.0

Note. X_{min}—minimum value of the characteristic, X_{max}—maximum value of the characteristic, X_{av}—the average value of the characteristic, S—mean quadratic derivation, V, %—coefficient of variation, S_x—mean arithmetic error, S_x, %—the relative mean arithmetic error, pcs.—pieces.

Thus, depending on the genotype of the plant, the length of the period from the beginning of the spring regrowth till the budding phase varied between 33 and 47 days, before blooming from 42 to 52 and ripening from 86 to 104 days. Differences in the height of the plants were 45 cm, varying in the range from 90 to 135 cm. Judging by the elements of the structure of the seed production, variation was even more significant, the coefficient of variation was higher than 20 percent. Particularly, strong variation was observed in the number and weight of the seeds per plant ($V = 29.8$–29.6%).

Accessing the relationship of the color of the flowers and vegetative organs, we were able to distinguish from the resulting diversity of the forms the following varieties of *G. orientalis*: white flowering with light-green leaves and stems; light-blue flowering with green and dark-green leaves and stems; blue flowering with green, dark-green and purple pigmentation of the leaves and stems; violet flowering with green, and dark-green and purple pigmentation of the leaves and stems.

The defined forms differed in size and shape of the leaves and bush. There were plants with small, medium, and large leaves, ovate, elliptical and lanceolate, with erect and semierect bush. The differences were also detected in height, the number of emerging stems, seed mass per plant, and mass of 1000 pieces (Table 26.2).

TABLE 26.2   Characteristic of the obtained plant forms of *Galega orientalis* found in the nursery of hybrid population of F_2

Color		Leaves		Form of bush	Height of plant (cm)	Stems on plant (pcs)	Mass of seeds per plant (g)	Mass of 1,000 seeds (g)
Flower	Leaves and stems	Size	Form					
White	Light-green	Medium	Ovate	Erect	105	10	22.1	7.3
Lilac	Green	Medium	Ovate	Erect	110	11	24.8	7.4
Blue	Green	Medium	Elliptical	Semi-erect	110	9	21.4	7.2
	Dark-green	Large	Ovate	Erect	92	14	36.8	7.5
	Green	Large	Lanceo-late	Erect	104	14	24.5	6.6
		Medium	Ovate	Erect	106	10	20.6	7.2
Blue	Dark-green	Medium	Oval	Erect	105	13	44.4	7.6
	Dark-green with violet pigmenta-tion	Small	Elliptical	Semi-erect	103	13	24.3	7.2
		Medium	Lanceo-late	Semi-erect	115	23	62.6	7.2
	Green	Medium	Ovate	Semi-erect	86	11	14.2	6.2
		Medium	Ovate	Semi-erect	90	24	62.7	6.6
Violet	Dark-green		Oval	Erect	122	14	35.4	7.2
		Large	Lanceo-late	Semi-erect	114	9	24.2	6.6
	With violet pigmenta-tion	Medium	Oval	Semi-erect	98	11	25.5	6.7
			Lanceo-late	Semi-erect	78	24	58.7	7.2
		Large	Ovate	Erect	105	4	13.3	6.4

TABLE 26.2 *(Continued)*

Color		Leaves		Form of bush	Height of plant (cm)	Stems on plant (pcs)	Mass of seeds per plant (g)	Mass of 1,000 seeds (g)
Flower	Leaves and stems	Size	Form					
	.Dark-green	Small	Lanceo-late	Semi-erect	110	15	16.2	7.4
Dark-violet	With in-tense violet pigmenta-tion	Small	Lanceo-late	Semi-erect	73	11	15.6	6.5
		Large	Ovate	Semi-erect	103	15	49.2	7.3
		Medium	Lanceo-late	Semi-erect	101	12	36.3	7,4

In the white flowering plants, for example, there were only light-green vegetative organs, medium-sized and egg-shaped leaves, and erect bush form. The average plant height was 105 cm, the number of the stems per plant is 10 pieces, and weight of the seeds per plant is 22.1 g, weight of 1000 seeds was 7.3 g. Among other phenotypes, blue flowering and violet flowering plants had the greatest variety of forms. As a result of the recombination of genes, they formed plants with green, dark-green and violet-pigmented leaves small, medium, and large in size, ovate, lanceolate, and oval in shape. As to the seed production, phenotypes with blue flowering and violet flowering with dark-green leaves were the best. They produced 62.6 and 62.7 g of seeds per each plant. As a result of the evaluation in the nursery of the second-generation hybrids, 82 plants with different manifestation of morphological and economically useful traits have been found. In order to obtain constant variety samples of each type, the best plants with white, lilac, light-blue, blue and violet flowering having complex of economically useful characteristics and properties were planted in isolated areas.

In isolation, as a result of self-pollination of each selected plant, they produced full-grown, with high-sowing quality seeds, and should be noted that the seed ovary in self-pollination was low, lot of beans were empty with no seeds, and the selected qualitative seeds were used for further propagation of constant samples and their comparative evaluation in the selection process. To do this, seedlings were grown from the seeds in the laboratory and they were planted in the fields. Each species was planted individually with the nutritional space of 70 × 70 cm². This has enabled us in the short term to propagate them and get the required amount of the seeds.

The result of the further studies was the formation of constant variety samples of *G. orientalis* with different combinations of morphological traits that have practical importance as standard variety samples for comparison of new varieties for their identification. Thus, by the color of flowers and vegetative organs, we have isolated and vegetatively propagated six standard variety samples (SEG) of different varieties. Among them are SEG-1—white flowers with light-green leaves and stems, SEG-2—lilac flowers with dark-green leaves and stems, SEG-3—with light-blue flowers and dark-green leaves and stems, SEG-4—with blue flowers and dark-green leaves and stems, SEG-5—with violet flowers and purple pigmented leaves and stems, SEG-6—with cream-colored flowers and dark-green leaves and stems.

The standard variety samples differed in plant height and seed production. The plant height in the standard variety samples ranged from 130 to 140 cm (Table 26.3).

TABLE 26.3   Seed production of standard variety samples of *Galega orientalis*

Standard variety	Height of plant (cm)	Stems per plant (pcs)	Number per one stem				Seeds per plant (g)	Seeds in pod (pcs)	Mass of 1,000 seeds (g)	Crude protein content (%)
			Trusses	Pods	Seeds (pcs)	(g)				
SEG-1	130	44	5.7	101.4	170.3	1.2	54.6	1.8	7.3	27.7
SEG-2	130	40	6.4	102.8	204.7	1.5	60.0	2.0	7.4	25.6
SEG-3	140	28	5.9	92.0	259.3	1.8	50.4	2.8	7.1	24.8
SEG-4	135	33	4.9	75.2	165.3	1.3	42.9	2.2	7.7	25.3
SEG-5	130	38	6.1	120.5	246.2	2.0	75.0	2.0	8.1	23.2
SEG-6	138	30	5.5	90.0	238.0	1.8	54.0	2.6	7.5	24.5
HCP_{05}	-						5.8	-		1.8

Note. Look denotations Table 26.1

The tallest (140 cm) was the variety sample SEG-3 with blue flowers, dark-green leaves, and stems. The number of the stems per plant in the third year of life depending on the variety sample was 28–44 pieces. The largest number (44 pcs) was observed in the white flower varieties SEG-1 with light-green leaves. According to the analysis of the structural elements of the seed production, it was found out that on the average one stem of the studied samples formed from 4.9 to 6.4 trusses ranging from 75 to 120 beans, from 165.3 to 259.3 pieces, mass from 1.2 to 2.0 g of seeds. The whole plant (bush) produced from 42.9 to 75.0 g of seeds, their mass of 1000 ranged from 7.3 to 8.1 g.

The average number of seeds per pod was 1.8–2.8 pieces and the empty pods were not excluded. Seed setting was the highest among light-blue flowering samples of SEG-3, and seed production in the violet flowering types of SEG-5 whose plant produced 75.0 g of seeds.

The crude protein content in dry matter of green mass in the phase of the beginning of budding in the studied samples was: in the white flower form 27.7 percent; in the lilac flower form 25.6 percent; in the light-blue flower form 24.8 percent; in the blue flower form 25.3 percent and in the violet flower form 23.2 percent. The highest protein content has a white flower form, indicating the prospect of using this trait in breeding.

Thus, standard variety samples are characterized by a certain set of quantitative and qualitative traits, making it possible for the identification of varieties [12, 13]. Their further development will continue. The established standard variety samples of *G. orientalis* have been used by us to develop a methodology for establishing the criteria of novelty, DUS of the test varieties for patentability.

The project of private methodology to test *G. orientalis* for DUS was developed at the Department of Breeding and Genetics at the Belarussian State Agricultural Academy and was tested on standard variety samples at the GVTS "Gorki variety testing station" and in 2013 was transferred to the state institution "State Inspection for Testing and Protection of Plant Varieties" for implementation into practice of state testing varieties of *G. orientalis* for identification and patentability. In other countries, members of the International Union of States on Protection of New Varieties up to date, there is a lack of standard variety samples of *G. orientalis* and methodology determining the criteria of novelty, DUS.

It allowed Belarus to contribute to the solution of this problem.

26.2 CONCLUSIONS

- The laws of manifestation of genetic variation in the hybrid generations of legumes can be effectively used in breeding *G. orientalis* to accelerate the breeding process and predict in hybrid populations planned genotypes differing in morphological and economically valuable traits.
- Effective source for obtaining genotypic variation in *G. orientalis* can be mutant white flowering plants obtained from perennial crops of the blue flowering population of the variety Nesterka as well as artificial hybrid populations derived from the crosses between selected parental forms.
- The law of homologous series in hereditary variation and found in it parallelism of the inheritance of traits in legumes served as a theoretical basis for accelerating morphogenetic process in *G. orientalis* to expand its range of genotypic variation.
- As a result of the selection work done by us in a short time constant variety samples of *G. orientalis* were created with genotypic variability characterized by differently colored flowers and vegetative organs. Among them are: SEG-1—white flowers with light-green leaves and stems, SEG-2—lilac flowers with dark-green leaves and stems, SEG-3—with light-blue flowers and dark-green leaves and stems, SEG-4—with blue flowers and dark-green leaves and stems, SEG-5—with violet flowers and purple pigmented leaves and stems, and SEG-6—with cream-colored flowers and dark-green leaves and stems.
- Variety sample SEG-5 with violet flowers had the best indicators of seed production. One plant produced 75.0 g of high-quality seeds.
- Variety sample SEG-1 with white flowers and light-green leaves and stems was the best as to crude protein content. This parameter compared to the other types was higher by 2.1–4.5 percent and made up 27.7 percent.
- The established variety samples of *G. orientalis* have been used as standard variety samples for developing a project of private methodology on variety testing for DUS.

KEYWORDS

- **Crop yield**
- *Galega orientalis*
- **Lupine**
- **Legumes**
- **Selection**

REFERENCES

1. Myasnikovich, M. V.; Lesnikovich, A. I.; and Dedkov, S. M.; Science of Belarus at the M Kartel, N. A.; and Khotyleva, L. V.; odern Stage. Minsk: Publishing House Beljrusskaja Nauka; **2006,** 213p. (in Russian).
2. Kartel, N. A.; and Khotyleva, L. V.; Law of the Republic of Belarus "On Patents on Plant varieties." Minsk; **1995,** 20 p. (in Russian).
3. Kartel, N. A.; and Khotyleva, L. V.; Law of the Republic of Belarus On the Entry of the Republic of Belarus in the International Convention on the Guard of New Varieties of Plants. Minsk: State Register of Cultivars and Arboreal-Shrub Breeds; **2006,** 96–97. (in Russian).
4. Kartel, N. A.; and Khotyleva, L. V.; Character and Terms of Protection of Plant Varieties in Accordance with UPOV Convention. Geneva: UPOV; **1997,** 110 p. (in Russian).
5. Kartel, N. A.; and Khotyleva, L. V.; Genetics and selection of plants. Science to the national economy. Minsk: Analytical Center NAS of Belarus; **2002,** C. 742–750. (in Russian).
6. Gusakov, V. G.; and Ganush, G. I.; Agrarian science—to the production. Science to the National Economy. Minsk: Analytical center NAS of Belarus; **2002,** 897–907. (in Russian).
7. Vavilov, N. I.; Theoretical Bases of Selection. Moscow: Nauka; **1987,** 512 p. (in Russian).
8. Genetics of cultural plants: leguminous, vegetable, water-melon. All Union Academy of Agricultural Sciences Named After Lenin, V. I.; Reds. Fadeeva, T. S.; Burenin, V. I.; Leningrad: Agropromizdat. **1990,** 287 p. (in Russian).
8. Bushuyeva, V. I.; Evolution and modern classification of angustifoliate lupin. Collection of scientific works of Belarussian agricultural academy. Selection of grain and leguminous crops on stability of the productivity, immunity and quality of grain. Bushuyeva, V. I.; Latypov, A. Z.; et al. Gorki; **1996,** 11–17. (in Russian).
9. Bushuyeva, V. I.; and Taranukho, G. I.; Selection of angustifoliate lupine on the complex of economically useful traits. Selection of scientific works of Belarussian agricultural academy Selection Genetic Methods of Increase of the Productivity of Agricultural Crops. Gorki; **1997,** 16–18. (in Russian).
10. Bushuyeva, V. I.; Using of new gene pool for creation of source material and varieties of angustifoliate lupine. Abstract of Thesis of Candidate's Dissertation. Zhodino; **1989,** 22 p. (in Russian).
11. Bushuyeva, V. I.; *Galega Orientalis*: Monography. Minsk: Ecoperspectiva; **2008,** 176 p. (in Russian).
12. Bushuyeva, V. I.; *Galega orientalis*: Monography. 2nd edition. (Compiled by Bushuyeva, V. I.; and Taranukho, G. I.) Minsk: Ecoperspectiva; 2009, 204 p. (in Russian).

CHAPTER 27

BIOCHEMICAL EVALUATION OF THE VARIETY SAMPLES OF THE RED CLOVER AND *GALEGA ORIENTALIS*

VERA IV. BUSHUYEVA

CONTENTS

27.1 INTRODUCTION

Red clover and *Galega orientalis* are perennial legume grasses widely used in feed production in the Republic of Belarus. They play an important role in providing animals with nutritious feed of high quality and obtaining livestock products with the lowest cost. Different types of feed are prepared from them: hay, haylage, silage, and green feed. With a variety of feeds, they can be included in the diet of the animals throughout the year. During the growing period, the grass stand of a high yielding clover and *G. orientalis* can ensure the production per hectare of 1.5–3.0 t of nutritious by fraction and amino acid composition protein, which is characterized not only by the high digestibility of itself, but also increases the digestibility of feeds from other crops. The low cost of feed produced from red clover and *G. orientalis* is provided by their ability as legumes to synthesize protein at the expense of the biological fixation of atmospheric nitrogen, while excluding the costs of energy-intensive and expensive nitrogen fertilizers [1]. However, the nutritional value of feed and the quality are largely dependent on the varieties of crops. Therefore, in the BSAA at the Department of Breeding and Genetics for many years, there have been breeding work on creating a new raw material and varieties of red clover and eastern *G. orientalis* with high yield and forage nutritive value. As a result of selective breeding varieties of red clover Mereya, TOS-870, and *G. orientalis* Nesterka were created, which are widely cultivated in the production of the Republic of Belarus. In addition, a new original material with a different combination of morphological and agronomic and biological characteristics and properties was produced [2–4]. To increase the nutritional value in the produced varieties, biochemical assessment of the initial material is an integral part of the selection process.

27.2 ANALYSIS OF LITERATURE

In the breeding of forage crops, one of the priorities is to improve the quality of feed, which is done mainly by increasing the protein content of essential amino acids, fat, ash, nitrogen-free extractives (NFEs) in the green mass, as well as reducing the amount of fiber. Fundamental studies of the chemical composition of the forage crops (collections in the All-Russian Institute of Plant Growing) were held by N.N. Ivanov at the dawn of biochemical science. They showed that all species and varieties of plants are characterized by the chemical variability [5]. For a successful breeding, it is important to know the factors that cause this or that quality variability and the limits within which it is manifested in different crops and varieties. During numerous studies [6–9], it was found that the amplitude of the quantitative changes in the biochemical composition and nutritive value of legume grasses depends on the phase of plant development. As you progress through the phases of vegetation, especially the flowering phase, protein content begins to decrease, fiber content increases, and the energy value of feed dramatically decreases. The chemical composition of forage varies by phase of development not only quantitatively but also qualitatively. This is due to the variability of the biochemical processes that occur in plants during the growing season. In the early phases of development, plants contain more nitrogen substances, easily soluble carbohydrates, and ash elements. During this period of the plant, growth respiration rate, the production of ATP in the cells, and the activity of redox enzymes are very high. Quantity of proteins in the leaves increases depending on their growth and increase of their area. When leaf growth ceases, the protein content in them remains constant for some time, and at the later development phases, and particularly during the formation of generative organs, there is a decrease of nitrogen and mineral content and increase of the amount of sparingly soluble carbohydrates and lignin. In this regard, by the end of the flowering phase and especially after flowering, digestibility and feeding value of the vegetative mass deteriorate. In the earlier phases of development, plants have a lot of proteins, mineral elements, an increased amount of fat, but fiber content is very low. In the process of growth and development of plants, they contain less protein, minerals, fats, and dramatically increased amount of fiber. The concentration of digestible carbohydrates in the vegetative organs of legumes increases to the budding phase, and then decreases. Significant changes are observed at the same time in the composition of nitrogenous substances. In young plants, crude protein contains rather many proteins, and their content decreases in the process of growth and the proportion of nonprotein nitrogen compounds increases. Young plants are characterized, as a rule, by high content of albumin, and at later stages of development, more sparingly soluble proteins predominate in plants. Significant changes in carbohydrate complex also take place. At the early stages of growth of legumes, there is much sugar and starch in the composition of NFEs, and after flowering, amount of sugars decreases and increases the concentration of hemicellulose. The ash content of grasses during the growing season decreases. Under optimum conditions, the nutritional content of calcium and magnesium in them at this time is almost unchanged, and that of potassium and especially phosphorus reduces significantly. Therefore, in the later stages of plant development, ratio of Ca:P and Ca:K in the ash increases dramatically. When feeding animals on the feed mass produced at late grass cutting, they may be lack in phosphorus. The content of almost all vitamins in the green mass during vegetation is also reduced. The limits of variability of each of the biochemical parameters are largely dependent on genetics, of both crop and variety. Therefore, in the selection process, study of biochemical characteristics of the breeding material and the influence of the phase of plant development on the productivity and quality of forage is the basis of success in the creation of varieties with high yield and forage nutritive value.

The aim of our study was to perform biochemical evaluation of the best variety samples of the red clover and *G. orientalis* and assess their nutritional value.

27.3 MATERIALS AND METHODOLOGY

The objects of study were red clover varieties Mereya, TOS-870, variety samples SL-38, Mut-BSAA, variety of *G. orientalis* Nesterka, and varieties SEG-1, SEG-2, and SEG-4 produced by the Department of Breeding and Genetics at the BSAA.

We studied the chemical composition of the green mass of the varieties of *G. orientalis* Nesterka and red clover Mereya in the phases of stemming, budding, and flowering, determining the variability of the content of protein, fat, ash, fiber, and NFEs, depending on the phase of the plant development.

Top variety samples of the red clover TOS-870, SL-38, Mut-BSAA, and *G. orientalis* SEG-1, SEG-2, and SEG-4 were analyzed by biochemical composition of green mass in the bud stage, early flowering. Red clover Mereya and *G. orientalis* Nesterka were used as the control cultivars.

The amino acid composition of proteins (% of total protein) in the green mass of the red clover varieties TOS-870 and variety samples of *G. orientalis* SEG-1, SEG-2, and SEG-4 were also studied.

Evaluation of the feed nutrition of the red clover and *G. orientalis* was carried out in the Chemical and Environmental Laboratory of the BSAA. The amino acid composition of proteins was determined by amino acid analyzer T339 (Czech Republic) [10].

27.4   RESULTS AND DISCUSSION

In the zoned varieties of *G. orientalis* Nesterka and red clover Mereya, we determined the content of protein, fat, ash, fiber, and NFEs in the phases of stemming, budding, and flowering. It was found that the highest content of protein, fat, and ash in the green mass of both crops were in the phase of stemming and the lowest in the phase of full bloom. Thus, reduction of quantitative parameters in *G. orientalis* beginning from the phase of stemming and ending at the phase of full flowering occurred in the following order: the protein content decreased from 27.6 to 18.5 percent, fat from 3.0 to 2.6 percent, and ash from 10.6 to 7.2 percent (Table 27.1).

TABLE 27.1   Chemical composition (% of completely dry matter) of green mass of *Galega orientalis* and red clover depending on the phase of development of herbage

Crop variety	Vegetation phase	Protein	Fat	Ash	Fiber	NFE
Galega orientalis, variety Nesterka	Stemming	27.6	3.0	10.6	25.5	33.0
	Budding	19.5	2.8	7.9	26.4	43.2
	Flowering	18.5	2.6	7.2	33.1	38.5
Red clover, variety Mereya	Stemming	21.4	2.9	10.4	22.2	32.4
	Budding	18.5	2.7	8.3	29.4	41.2
	Flowering	16.2	2.5	7.4	30.2	36.0

In the red clover, these figures were reduced in the same sequence. The highest protein content value (21.4%) was in the phase of stemming and the lowest (16.2%) in the flowering phase. In the budding phase, the figure was intermediate and was 18.5 percent. The fat content was reduced by the phases of the development from 2.9 to 2.5 percent, and ash from 10.4 to 7.4 percent. The content of the dietary fiber and NFEs, on the contrary, increased with the growth and development of plants. In the phase of plant stemming, their content was the lowest and amounted to 25.5 percent of cellulose and 33.0 percent of NFE in *G. orientalis* and in the red clover 22.2 and 32.4 percent, respectively. It should be noted that the maximum accumulation of NFEs occurred in the budding phase when its content was 43.2 percent in the *G. orientalis* and 41.2 percent in the clover. In the flowering phase, that figure fell to 38.5 percent in *G. orientalis* and 36.0 percent in the red clover. The fiber content increased to 33.1 percent in *G. orientalis* and 30.2 percent in the clover. The obtained results of studying the chemical composition of the green mass of both the crops in different phases of development of the plants were the only proof that the most optimal timing of harvesting forage is the phase of budding, which is the beginning of flowering. In this phase, the herbage of the both crops forms the highest yield of forage. Therefore, a more complete biochemical estimation of the feed mass of the variety samples of the red clover and *G. orientalis* was carried out by us in the budding phase, which is the beginning of flowering.

According to the previous studies [2–4], the average yield of green mass harvested in this phase was in the cultivar of *G. orientalis* Nesterka 650 cwt/ha, in the red clover varieties Mereya—570 cwt/ha. When you consider that 1 kg of green mass of *G. orientalis* contains on an average 0.25 fodder units and 45 g of digestible protein (DP), and in the red clover—0.2 fodder units, and 30 g of DP, then by cultivating varieties of *G. orientalis* Nesterka, we can obtain 162.5 cwt of fodder units or 29.2 cwt/ha of DP, and in the varieties of the red clover Mereya—114 cwt/ha of fodder units or 17.1 cwt/ha of DP. Thus, one feed unit of *G. orientalis* contains 176.0 g of DP and that of red clover 150.0 g. However, in the feed production in the Republic of Belarus, feed quality and protein diet supply still remain relevant. One feed unit accounts for only 99 g of digestible protein, which is below normal by 6 g. Deficiency of protein in animal feed, according to scientists, leads to an annual expenditure of more than 1 million tons of grain in the republic. Therefore, the cultivation and use in the feed production of the varieties of red clover Mereya and eastern *G. orientalis* Nesterka in which the content of digestible protein in one feed unit is 150–176 g certainly will improve the quality of feed and reduce the protein deficiency. Creating more productive varieties of the both crops with higher feed qualities will further enhance their effectiveness. It is known that depending on the type of the crop and genotypic characteristics of the variety, the nutrient content of the fodder mass may be different.

To identify these differences, we carried out a comparative biochemical evaluation of the new variety samples of the red clover TOS-870, SL-38, Mut-BSAA, and *G. orientalis* SEG-1, SEG-2, and SEG-3 with the varieties Mereya and Nesterka. The evaluation revealed differences between the variety samples for each crop. So, in the green mass of the variety samples of the red clover, dry matter content ranged from 21.42 to 22.60 percent, in *G. orientalis* from 20.3 to 24.0 percent (Table 27.2).

TABLE 27.2 The chemical composition (% to dry matter) of the green mass in the variety samples of *Galega orientalis* and red clover in the budding and early flowering phase

Trait	Red clover				Galega orientalis			
	Mereya control	TOS-870	SL-38	Mut-BSAA	Nesterka control	SEG-1	SEG-2	SEG-4
Dry matter, %	21.49	21.42	22.12	22.6	22.9	20.3	24.0	23.7

TABLE 27.2   *(Continued)*

| Trait | Red clover | | | | Galega orientalis | | | |
	Mereya control	TOS-870	SL-38	Mut-BSAA	Nest-erka control	SEG-1	SEG-2	SEG-4
Protein	16.32	16.42	18.38	16.55	16.06	19.23	18.76	17.01
Fat	2.95	3.00	3.40	3.21	2.99	3.14	3.03	3.06
Ash	11.25	11.28	11.20	12.30	11.06	11.20	11.12	11.06
Fiber	25.29	25.62	22.4	23.30	27.52	25.40	25.70	26.20
NFE	37.69	37.81	39.99	38.32	35.81	37.60	34.10	38.00
Calcium	1.29	1.21	1.36	1.29	1.33	1.21	1.28	1.30
Phosphorus	0.76	0.73	0.95	0.84	0.65	0.74	0.69	0.66
Sugar	4.05	4.10	4.50	4.20	4.00	4.64	4.31	4.23
Carotene (mg/kg)	34.0	34.5	37.8	35.2	33.50	38.11	36.5	34.9

The highest rate of dry matter content was found among variety samples of the red clover Mut-BSAA (22.6 vs 21.49% in the variety Mereya); in *G. orientalis*—in SEG-2 (24.0 vs 22.9% in the variety Nesterka). The differences between the variety samples of the two crops have also been identified by other indicators of the biochemical analysis. In the red clover, the limits of variability among variety samples were on protein content 16.32–18.38 percent; fat 2.95–3.40 percent, ash 11.20–12.30 percent; fiber 22.40–25.62 percent; NFEs 37.69–39.99 percent; Ca—1.21–1.36 percent; P—0.73–0.95 percent; sugar—4.05–4.50 percent; carotene—34.0–37.8 mg/kg. In *G. orientalis*, differences among samples on the biochemical composition were also revealed. The variation of biochemical parameters depending on the samples had the following limits: the protein content of 16.06–19.23 percent; fat 2.99–3.14 percent; ash 11.06–11.20 percent; fiber—25.40–27.52 percent; NFE 34.10–38.00 percent; Ca—1.21–1.33 percent; P—0.65–0.74 percent; sugar—4.00–4.64 percent; carotene—33.5–38.11 mg/kg. The comparison of the biochemical indices of the green mass between red clover and *G. orientalis* showed that there are differences between them, but both of them have high feeding value as to the nutrient content composition. As for the variety samples and their comparative assessment of each other, the variability of quantitative parameters of the biochemical composition that occurs among them has allowed us to make a selection of the samples exceeding the nutrient feeding value of the red clover Mereya and *G. orientalis* Nesterka. Among red clover plants to such variety samples belong SL-38 and Mut -BSAA, as to *G. orientalis*: SEG-1 and SEG-2. Along with the better feed nutrition, these variety samples in comparison with the control were characterized by higher production potential. The results of these studies showed that the biochemical composition and feeding value of the variety samples of red clover and *G. orientalis* depend on the age of the plant, as well as the specific and varietal characteristics of the crop. An important characteristic of the feed derived from the variety samples of red clover and *G. orientalis* is their biological nutritional value, which is determined by the amino acid composition of the protein. Proteins are more nutritious if they are balanced in amino acid composition and contain more essential amino acids. The biological nutritional value of the proteins in the studied variety samples was defined by us comparing them with the total percentage of the most valuable and highly digestible protein of chicken eggs, as well as with the theoretical protein, which is currently recommended by FAO as "ideal." In accordance with FAO, the content of essential amino acids in the theoretical protein should be equal to threonine 4.0 percent, valine 5.0 percent, methionine 1.7 percent, isoleucine, 4.0 percent, leucine 7.0 percent, phenylalanine 2.6 percent, lysine 5.5 percent, and tryptophan 1.0 percent and their sum is 30.8 percent (Table 27.3).

TABLE 27.3 Amino acid composition of the protein (% of total protein) in the green mass of red clover and *Galega orientalis*

Amino acid	Red clover TOS-870	Variety samples of *Galega orientalis*			Lucerne Daishi	Egg proteins	Standard FAO
		SEG-1	SEG-2	SEG-4			
Aspartic	12.49	13.88	12.72	15.13	12.7	4.3	-
Threonine	3.95	4.03	4.14	4.06	3.85	3.6	4.0
Serine	5.49	4.87	5.28	5.13	6.19	3.0	-
Glutamic	10.55	10.59	11.22	10.29	10.25	13.7	-
Proline	14,96	12.89	12.58	13.78	15.16	6.4	-
Cystine	0.11	0.28	0.09	0.11	0.09	0.8	-
Glycine	4.42	4.12	4.38	4.11	4.62	2.9	-
Alanine	5.35	5.10	5.31	5.10	5.45	5.8	-
Valine	5.38	4.98	4.90	4.69	6.88	5.5	5.0
Methionine	0.57	0.61	0.59	0.49	0.58	1.8	1.7
Isoleucine	3.52	3.43	3.41	3.29	3.50	5.7	4.0
Leucine	6.80	6.86	7.18	6.77	6.5	6.5	7.0
Tyrosine	3.85	3.82	3.79	3.35	4.25	2.2	-
Phenylalanine	4.29	4.59	4.65	4.40	3.59	3.3	2.6
Histidine	3.69	4.57	4.46	4.60	3.67	3.6	-
Lysine	5.19	5.61	5.81	5.52	5.79	8.5	5.5
Arginine	4.48	4.74	4.57	4.23	4.28	13.2	-
Tryptophan	1.10	1.20	1.20	1.15	1.20	1.3	1.0
Sum	96.19	95.67	95.78	95.80	98.55	92.1	-
Sum of essential amino acids	30.80	31.31	31.88	30.37	31.89	36.2	30.8

[1]In the egg protein, the sum of all amino acids is 92.1 percent, including essential 36.2 percent [1]. In our studies in the crude protein of the green mass of red clover and *G. orientalis*, 18 amino acids were identified, including all essential. In relative terms, that is in percentage to crude protein, all amino acid contents in the variety samples amounted to: in red clover TOS-870—96.19 percent; in *G. orientalis* SEG-1—95.67 percent, SEG-2—95.78 percent, SEG-4—95.80 percent; in alfalfa—98.55 percent against—92.1 percent in the egg protein (with its general content of 12.0%). Thus, the content of all amino acids in the protein of the studied samples was higher than in the egg protein by 3.57–4.09 percent. In alfalfa, this parameter was higher by 6.45 percent. To determine the biological nutritional value of proteins, we studied the overall content and quality composition of the essential amino acids. Our analysis of the proteins in the green mass of the variety samples of both crops showed that they contain all the essential amino acids and fairly well-balanced amino acid composition. The concentration of most of the essential amino acids in individual samples was higher than that required by the rules of feeding farm animals in accordance with the FAO standards. It should be noted that amino acid composition of the protein depended on the species and variety features. So, higher threonine content was observed in the samples of *G. orientalis* SEG-1 (4.03%), SEG-2 (4.14%), and SEG-4 (4.06%) versus 4.05 percent for the standard FAO and 3.6 percent for egg protein. The highest content of valine (5.38 vs 5.0% in the standard FAO) was observed in red clover samples TOS-870, and leucine in the samples of *G. orientalis* SEG-2 (7.18%) versus 7.0 percent in the standard FAO and 6.5 percent for egg protein. All samples of *G. orientalis* and red clover had higher content of lysine, tryptophan, and phenylalanine than the FAO standard. The highest amount of essential amino acids (31.88 and 31.31%) was observed in the samples of *G. orientalis* SEG-2 and SEG-1. The red clover variety in essential amino acid content was at the level of the FAO standard. It is known that animals absorb from plant feed essential amino acids. The lower the biological nutritional value of the plant protein, the higher is the feed consumption. It is believed that every missing gram of protein in the feed unit leads to excessive consumption of feed up to 2.0 percent. It is therefore very important to use for animal feeding nutritious protein feed. With balanced protein diet, energy costs and consumption of feed reduce and the productivity of the animals increases.

27.5 CONCLUSIONS

The creation of new high-yielding varieties of red clover and *G. orientalis* with improved quantitative and qualitative composition of the essential amino acids is one of the ways to successfully solve the problem of feed protein in the Republic of Belarus. Created by the Department of Breeding and Genetics, the new source material is characterized by high biological nutritional value. The best by the biological nutritional value of the protein among the studied variety samples were *G. orientalis* SEG-2, SEG-1, and red clover TOS-870 that exceed the FAO standard not only by the content of individual amino acids, but also by their total number.

KEYWORDS

- *Galega orientalis*
- **Green grass**
- **Lysine**
- **Phenylalanine**
- **Red clover**
- **Tryptophan**

REFERENCES

1. Bushuyeva, V. I.; Physiology and biochemistry of farm crops. Study Manual for Higher Institutes. (Compiled by Tretyakov, N. N. et al.,) 2nd edition. Moscow: Kolos; **2005**, 656 p. (in Russian).
2. Bushuyeva, V. I.; New Variety of the Red Clover Mereya. Bushuyeva, V. I.; Bulletin of the Belorussian State Agricultural Academy; **2006**, *2*, 51–55. (in Russian).
3. Bushuyeva, V. I.; New Variety of *Galega orientalis.* Nesterka: Bulletin of the Belorussian State Agricultural Academy; **2006**, *3*, 50–56 p. (in Russian).
4. Bushuyeva, V. I.; *Galega orientalis*. Monography. 2nd edition. (Compiled by Bushuyeva, V. I.; and Taranukho, G. I.;). Minsk: Ecoperspectiva; **2009**, 204 p. (in Russian).
5. Ivanov, N. N.; Biochemical basis of the plant selection. Ivanov, N. N.; Theoretical Basis of Selection. Moscow: Leningrad; **1935**, *1*, 991–1016. (in Russian).
6. Bekuzarova, S. A.; Selection of the red clover. Bekuzarova, S. A.; Monography. Vladikavkaz: Gorski State Agricultural University; **2006**, 176 p. (in Russian).
7. Novoselova, A. S.; Selection and Seed Growing of Perennial Grasses. Moscow: Publisher Named After Bolkhovitinov, E. A. **2005**, 375 p. (in Russian).
8. Kshnikatkina, A. N.; *Galega orientalis.* Monography. Penza: RIO PGSHA; **2001**, 287 p. (in Russian).
9. Truzina, L. A.; *Galega orientalis*: the history of research and technological bases of its cultivation in the non-black soil zone. Truzina, L. A.; Mosin, S. V.; Feed Production; Problem and Ways of Solution. Moscow: FGNU Rosinformagrotekh; **2007**, 164–172 p. (in Russian).
10. Pleshkov, B. P.; Manual on Biochemistry of Plants. Moscow: Kolos; **1985**, 255 p. (in Russian).

INDEX

Milton Keynes UK
Ingram Content Group UK Ltd.
UKHW050256161024
449569UK00042B/1735